# Contents

# Preface

One hundred years after the first patent on arc welding was filed, the public image of welding is still that of a man behind a mask, creating bright lights, smoke, and sparks. For the TV viewer, this is the spectacular scene which is used to evoke the idea of heavy industry. For others, welding is associated with a garage mechanic repairing a damaged or rusted wing panel on a car. Even the trained engineer often has a restricted view of welding, limited to a large extent by everyday experience of one company's practices. Yet the term 'welding' covers a large number of techniques which can be used by the production engineer. Indeed, it is this very diversity which has made welding indispensable to manufacturing industry. So many of the things we rely on for modern living would not be practicable were it not for welding — from the construction of oil platforms and power plant, through volume production of cars, to the sealing of transistor capsules. It is therefore surprising, to say the least, that the technology of welding is so poorly understood by many engineers, technicians, and craftsmen. The effect of this is to be seen in poor-quality products and lost opportunities for significant improvements in productivity. In writing this book, I have tried to make some small contribution towards remedying this situation by providing an insight into the technology and by highlighting the various options open to the production engineer.

One of the problems of teaching welding technology at an introductory level is that it involves an unavoidable amount of description which can easily lose the student's interest. I have tried to overcome this problem by concentrating on the logic behind the existence of the various techniques. The book is intended to be read as a continuous text, and I hope that in doing so the reader will capture some of the intellectual challenge which comes from the involvement of many basic disciplines. This type of treatment has meant that the book does not follow any particular examination syllabus, although it does cover the content of the Business & Technician Education Council units on welding and of the City and Guilds craft courses. It has also been used on more general engineering courses at undergraduate and technician level.

In the preface to the first edition, I wrote that welding is a developing technology. The intervening years have served to confirm this, and many techniques which were then at the development stage are now standard practice. The issue of a second edition offers an opportunity to introduce new material. In particular, I have described the use of solid-state and inverter systems in the construction of power-supply units. I have also included an introduction to the role of robots in welding operations, and I have brought the section on laser welding up to date.

I would like to thank my many colleagues who have made helpful comments on the text and to acknowledge the help given by The Welding Institute with many of the illustrations. I would also like to thank the following organisations for supplying illustrations of items of their equipment: BOC Ltd (figs B1(b) and B4), ESAB (UK) Ltd (figs 4.24, 10.6, 10.21, and 10.24), and Thompson Welding Systems (fig 10.19). Material from British Standards BS 499:part 1:1983, BS 639:1976, BS 679:1959, BS 2573: part 1:1983, and BS 5500:1985 has been used by permission of the British Standards Institution, 2 Park St, London W1A 2BS, from whom copies of the complete standards may be obtained. Finally, I want to record my continuing thanks to Bob Davenport of Edward Arnold (Publishers) Ltd for his patient and invaluable help in preparing the text for printing.

L.M. Gourd

# 1    Which method of joining?

## 1.1  The development of joining techniques

Though metals have been used by man for many thousands of years, no one is certain how the first useful metal was produced. It may be that metal nuggets left on the surface of the earth by meteorites were found to have useful properties. More likely, an early inhabitant of an area containing copper-bearing minerals inadvertently heated these ores in a open fire. Under suitable conditions this would have produced a lump of impure copper which could be hammered into shape. Whatever their origins, the early use of metals has been confirmed by the discovery of implements made from bronze. Axes, spearheads, and ornaments have been excavated from the sites of primitive settlements, and archaeologists have been able to show that they were made and used in a period which we now call the BronzeAge.

Bronze is an alloy of copper and tin, and one of its characteristics is that it can be shaped to give a good cutting edge. This would have made it very attractive to the late Stone Age hunter, who had been forced to do the best he could with flint axes and spears. Also, the introduction of bronze into domestic tools meant that constructional techniques, especially with wood, were improved.

The uses to which the newly found metal could be put were limited by the fact that the technology available could not offer techniques for producing large items entirely in bronze. This was not a handicap in the case of an axe or a spear, since there were operating reasons for using a resilient material such as wood for the shaft. Various methods were devised for joining the head to the shaft which satisfied the requirements of the time, but the problem of achieving acceptable metal-to-metal joints was to remain unsolved. Indeed, apart from the development of forge welding, which is normally attributed to the Syrians, about 1400 BC, the inability to join small pieces of metal to produce a larger or more complex component hampered engineering progress even up to the last century. The developments that did take place were largely in response to military requirements. A good example is the riveting of suits of armour, which enabled damaged panels to be replaced effectively. It needed the industrial revolution, with its dependence on machinery, to provide the incentive which led to the introduction on a commercial scale of bolting, riveting, soldering, and finally welding.

Today there are a large number of joining techniques available, and the problem is not how to join but how to select the best method of joining. Whereas bronze-age man simply had to chose between thongs and wedges, a design engineer today may well find that four or five techniques appear to be equally suitable. Each method has its own attributes, and a number of aspects must be evaluated if the final choice is to be sensible. The relative importance of such factors as strength, ease of manufacture, cost, permanency, corrosion-resistance, and appearance depends very much on the specific application.

The designer of a bridge looks for a means of producing joints in plates which will carry the varying loads imposed by vehicles travelling across the deck. While it would obviously be desirable to carry out the joining operations as quickly as possible, this might not be a prime consideration in all cases – the fact that some of the joints will have to be made on site could well be an over-riding factor. Thus, even though fillet welding is widely used in fabricating girders, critical site joints connected with these girders may be bolted in the interests of better control of quality and ease of working in difficult locations (fig. 1.1).

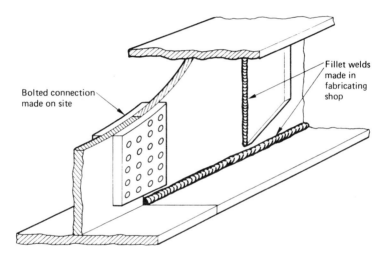

**Fig. 1.1**    Fillet-welded girder with bolted web connection

By contrast, the manufacturer of cars travelling across the bridge will be looking for joining techniques which can be incorporated into production lines, combining speed with reproducibility and reliability. As far as possible they must be capable of being used by semi-skilled operators, since a large labour force is usually required. This, in turn, demands that the production sequence is established and controlled or monitored by specialist personnel.

## 1.2 Manufacturing methods
Any discussion of joining presupposes that the decision to assemble the unit from a number of parts is correct. This may not always be a valid assumption since there are many viable alternatives. Casting, forging, pressing, bending, extrusion, spinning, and machining can all be used to produce both large and small components in a variety of metals.

These processes are by no means the only ones available, but they illustrate the range of manufacturing techniques on offer to the modern engineer. It is important to recognise that they are not mutually exclusive, and an economic production sequence will normally incorporate two or more of these methods linked to the judicious use of suitable joining techniques. We must also be careful to see joining – especially welding – not as something existing in its own right but as part of a corporate activity that takes pieces of metal and converts them into the product which the customer wants. It is also wrong to think only of sheets and plates when considering joining – frequently we are faced with the need to produce connections to or between castings, forgings, or extruded sections (fig. 1.2). This must be borne in mind as we review the various joining systems and subsequently discuss welding in detail.

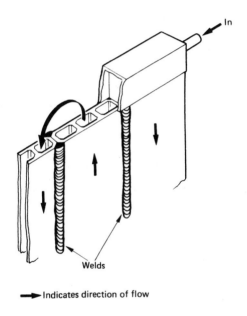

Fig. 1.2   Cooling panel made by welding together aluminium extrusion

## 1.3 Mechanically formed joints
Simple butt and corner joints can be made by folding and interlocking the edges of sheet-metal components (fig. 1.3). A large amount of deformation is needed to produce a secure joint, and this method is applicable only to material which has good ductility and can be bent around small radii. The strength of the joints is acceptable for many applications such as the manu-facture of sheet-metal containers and ducting. Where higher strength or better leak-tightness is required, the folds can be filled with solder, as in food cans.

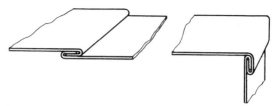

**Fig. 1.3**   Mechanical lock joints

## 1.4 Lap joints

Lap joints are used in most units fabricated from sheet metal (fig. 1.4); they are less frequently used to join plates. There are three basic types – butt, 'T', and corner – each of which can be joined in two different ways: we can either connect the sheets only at points along the length of the joint or provide a continuous bond over the complete interface.

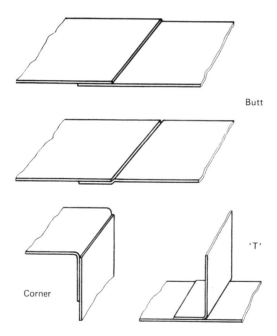

Butt

'T'

Corner

**Fig. 1.4**   Types of lap joint

Various methods are used to produce a local connection between the two parts. In structural work, bolted lap joints are common since they are easy to assemble on site or in the shop, provided the holes have been accurately drilled. Also, if high-strength friction-grip bolts are used, appreciable loads can be transmitted.

Another common technique used to secure lap joints is riveting, which was at one time the mainstay of the shipbuilding industry (it has now been replaced by welding). Rivets are widely used in sheet work as they readily give a joint of good strength and acceptable appearance. The skin panels on an aircraft contain many examples of the skilful use of rivets to provide joints which can withstand high stress. Blind or pop rivets are of particular interest where access is limited to one side of the joint.

If the thickness of the material is less than about 2 mm, local connections can be made at the interface by fusing slugs of metal (fig. 1.5). The most common example of this technique is resistance spot welding, which is described in detail in chapter 11.

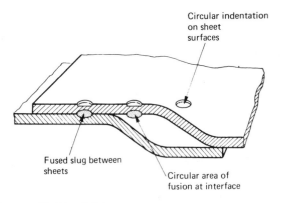

**Fig. 1.5**    Spot-welded lap joint

Both riveting and spot welding produce permanent connections, whereas bolting enables the joint to be dismantled for repair or modification. Bolting tends to be slow for sheet-metal work and is frequently replaced by the use of self-tapping screws or threaded fasteners.

None of the techniques so far discussed gives a leak-tight joint without the use of a sealant. This problem is eliminated if we can bond the sheets together across the complete area of the overlap; at the same time, the amount of overlap can be reduced for the same strength. There are three possible ways of achieving this: soldering, brazing, and adhesive bonding.

**Soldering**    In soldering, a small gap between the sheets at the overlap is filled with a low-melting-point lead–tin alloy known as a solder (fig. 1.6).

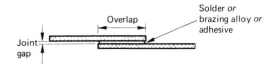

**Fig. 1.6**    Lap joint for soldering, brazing, or adhesive bonding

5

If the surfaces have been cleaned with a flux, the solder wets the metal and produces an intermetallic bond. If the gap is controlled to about 0.08 mm, the solder is drawn into the joint by capillary action, to give uniform filling. The melting point of the solder ranges from 183 to 275°C, depending on composition, and the joint is heated to the appropriate temperature by a soldering iron or gas flame.

**Brazing** is similar to soldering but uses fillers having higher melting points (450 to 800°C). Brazed joints usually have better strength than soldered ones, but the need to heat to higher temperatures can pose problems, especially as oxidation and discoloration of the metal surfaces can occur. Various heating methods are used, such as gas torch, furnace, molten-flux bath, induction heating, and resistance heating.

**Adhesive bonding**   Compared with soldering and brazing, this is a relatively new technique, but it has found rapid acceptance due to its ease of use. As with soldering and brazing, a small gap is used which in this case is filled with adhesive which forms a surface bond with the metal. The type of adhesive is varied to suit the particular requirements of the joint.

A lap joint loaded in tension rarely experiences simple shear (fig. 1.7) – the applied forces are not in line but are displaced, with the result that the joint tends to rotate as the forces try to achieve alignment. The use of butt straps enables the sheets or plates to be placed edge to edge so that the applied forces are acting in one plane. Such joints are more expensive to produce, require more material, and are heavier than the simple lap. In the majority of cases, a welded joint offers a better solution by providing a direct connection between the abutting edges of the parent metals.

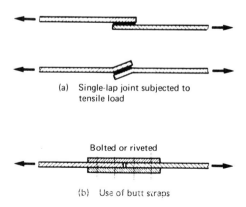

(a)   Single-lap joint subjected to
tensile load

Bolted or riveted

(b)   Use of butt straps

**Fig. 1.7**   Lap joints under load

## 1.5 Welded joints

Most welded joints are made by melting the parent material on each side of
the joint lines (fig, 1.8). The molten metal combines to form a liquid pool
between the two components (if necessary, additional metal is added to build
up the cross-section of the weld). When the pool solidifies, a continuous
metallic bridge is produced which is able to carry loads and is leak-tight.

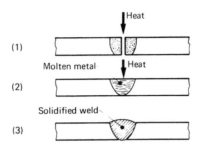

**Fig. 1.8**   Formation of a fusion-welded joint

Unlike the other methods of joining discussed so far, the success of fusion
welding depends very much on the skill of the person operating the process.
Bolting, riveting, soldering, brazing, and indeed spot welding involve operations which can be standardised and adapted to machine production and
control. While this can sometimes be done with fusion welding, a number of
operating factors rule out the widespread use of automatic welding. This is
not necessarily a disadvantage, however, since the craft skill possessed by the
welder provides a number of versatile techniques which can be used to
fabricate products ranging from small-bore thin-walled tube systems to ships
and bridges. Of equal importance is the ability to accommodate one-off
fabrications and to vary the type of work without having to invest in a major
re-equipment programme.

Although it is true to say that many engineering advances such as the jet
engine and the generation of nuclear power would not have been possible
without welding, it is only one of a number of joining methods. Various
advantages can be claimed for its use, but it must be judged on its merits
in comparison with the other techniques we have discussed. Welding should
be used only when it is the cheapest method of fabrication or when it offers
special technical advantages. As a technology, welding has progressed rapidly
in the last three decades, and we need to understand the principles involved
if it is to be used effectively. In this book, we will be exploring the background
to welding, the way in which the processes work, and the factors that govern
the production of an acceptable weld.

# 2 Methods of welding

## 2.1 Types of welded joint

It is often extremely difficult to give accurate but brief definitions of common everyday things. This is very true of welding — daily we use equipments which rely on welded joints for their satisfactory operation, but rarely do we stop

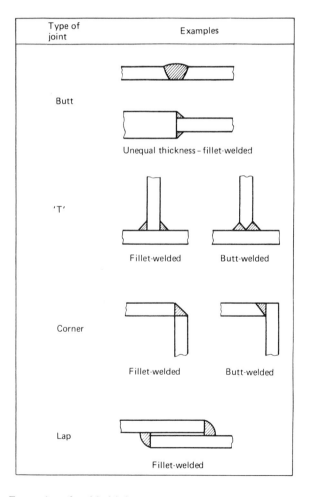

Fig. 2.1  Examples of welded joints

to consider the essential characteristics of the welds themselves. If we did, it would immediately become apparent that there are very many types of weld, which seem to vary according to the type of joint used. However, closer investigation would reveal that both the welds and the joints can be categorised into groups. We would discover that there are four basic types of joint: butt, 'T', corner, and lap (fig. 2.1).

The butt joint is characterised by the fact that the edges of the components are abutted and the load is transmitted along the common axis. This joint is of particular importance in fabrication, being used, for example, to join lengths of pipe, plates in ships' hulls, and flanges on bridge girders (fig. 2.2).

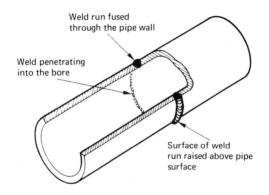

Weld run fused
through the pipe wall

Weld penetrating
into the bore

Surface of weld
run raised above pipe
surface

**Fig. 2.2**    Butt weld between two lengths of pipe (sectioned to show bore)

The 'T' joint is probably the most commonly used connection in fabrication. Typical examples are at the flange-to-web junction in a plate girder, stiffeners welded to a panel, branches attached to a main pipe, and lifting lugs (fig. 2.3). The joint can be made either with no penetration along the joint line (using 'fillets' of weld metal to provide the load-carrying connection) or with bonding across the interface, i.e. butt-welded.

The corner joint can similarly be butt- or fillet-welded according to service requirements. Corner joints are normally associated with box sections, but it is worth noting that, from the point of view of both design and production, flange-to-pipe connections represent very important examples of corner joints.

Finally, the lap offers an interesting variant because it can be used in butt, 'T', and corner joints. Most commonly, the bonding in a lap joint is over only a small area of the interface, either as a number of spots (fig. 2.4) or as a narrow strip along the length of the joint. Lap joints are mainly used in sheet fabrication, and common items such as cars, washing machines, refrigerators, etc. contain many examples.

An important feature of all the joints discussed above is the need for a bond which will transmit loads. This means that between the two components

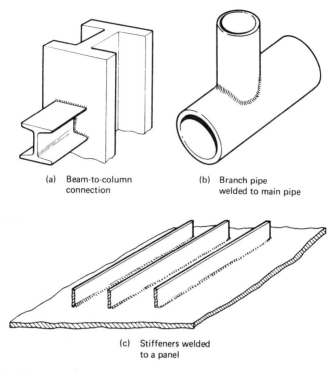

(a) Beam-to-column
    connection

(b) Branch pipe
    welded to main pipe

(c) Stiffeners welded
    to a panel

**Fig. 2.3** Examples of 'T' joints which have been fillet-welded

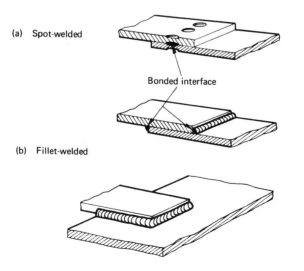

(a) Spot-welded

Bonded interface

(b) Fillet-welded

**Fig. 2.4** Examples of lap joints

there must be a continuous metallic connection – i.e. a metallic bridge – which will be strong enough to withstand the stresses applied in service.

## 2.2 Weld formation

We are now in a position to consider the definition of a weld, and in particular to look at the definition given in British Standard 499:part 1:1983, 'Welding, brazing and thermal cutting glossary':

> 'A weld is a union between pieces of metal at faces rendered plastic or liquid by heat or by pressure or by both.'

The first part of this definition follows from our discussion above: the union is the continuous load-bearing metallic bridge to which we have referred. But what is the significance of the reference to rendering the faces plastic (i.e. permanently deformed) or liquid? To understand this we must first consider how the union is achieved.

Sufficient strength can be produced in a welded joint only by interatomic bonding, and the prime function of the welding operation is, therefore, to provide links between atoms at the interface of the joint. For these links to be formed, two conditions must be satisfied. Firstly, the surfaces must be in *intimate contact*. This implies that they should be atomically flat so that when they are brought together the gap between the respective surface atoms will be about equal to the atomic spacing within the metal, i.e. $1.24 \times 10^{-10}$m. Even polishing to a high finish with diamond dust is unlikely to produce this degree of flatness. Secondly, the surfaces must be metallurgically clean – any molecules of grease, paint, moisture, oxygen, or nitrogen present on the surface will prevent the metal atoms from uniting, even if intimate contact is achieved.

Hence, if a practical welding system is to be based on the idea of simple interatomic bonding, a means of bringing the surfaces into intimate contact and, at the same time, dispersing the surface contaminants is essential.

## 2.3 Cold pressure welding

One way of achieving intimate contact and dispersal of contaminants is to force the surfaces together (fig. 2.5). Under pressure the surfaces deform, breaking up the contaminants and bringing areas of clean metal into intimate contact.

The pressure required to disperse the contaminants leads to appreciable reductions in the thickness of the workpieces. Optimum joint strength is obtained at a level known as the threshold deformation, the actual value of which depends on the metals being welded (Table 2.1): in general, the softer the metal, the lower the deformation required to initiate welding at room temperature.

Cold pressure welding is used to a limited extent to make welds between aluminium cable and connectors and for specialised applications such as welding caps to tubes (fig. 2.6), but it is usually difficult to accommodate the amount of deformation required for the welding of commercial alloys.

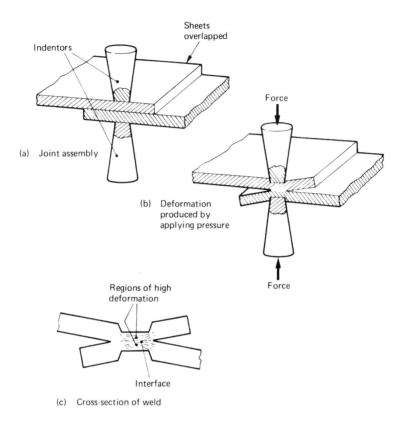

(a) Joint assembly

(b) Deformation produced by applying pressure

(c) Cross-section of weld

**Fig. 2.5** Cold pressure spot welding

**Table 2.1** Threshold deformations for welding

| Metal | Threshold deformation (% reduction in thickness) |
|---|---|
| Lead | 10 |
| Tin | 15 |
| Aluminium | 40 |
| Copper | 45 |
| Iron | 65 |

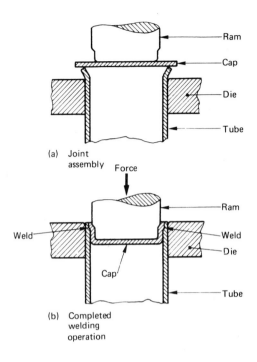

(a)   Joint assembly

Force

(b)   Completed welding operation

**Fig. 2.6**   Fitting caps to tubes by cold pressure welding

## 2.4 Hot pressure welding

Greater success can be achieved if the metals are heated during the welding operation. Raising the temperature reduces the value of the threshold deformation, and a number of successful hot-pressure-welding techniques have developed. Probably the oldest of these is forge welding, which has been in use by blacksmiths since about the year 1400 BC. In this process, the wrought iron or steel bars which are to be joined are heated to about 1350°C. At this temperature the iron oxides on the surface are melted and, when the components are hammered together, the molten oxides are squeezed out of the joint. Bonding then occurs at relatively low deformation levels.

Forge welding is now mainly used for craft work, but the principle of heating the component to make it easier to achieve a weld is also used in a number of modern developments of hot pressure welding. The method of heating is not critical, since the prime object is to raise the interfaces to a temperature at which the threshold value falls to about 25%. Since the hot metal is more plastic than it would be in cold pressure welding, the force which is required is smaller and it is possible to produce welds in hard metals such as steel. The resulting joint shows a similar structure to that of the cold pressure weld, but there is no evidence of cold working, since the deformation takes place above the recrystallisation temperature (see page 102).

13

Of the various methods used to heat a joint for hot pressure welding, three of the most successful are gas heating, resistance heating, and induction heating.

**Gas heating**   A ring burner is arranged around the joint, and mixed oxygen and acetylene gas is supplied to burners spaced uniformly around the circumference (fig. 2.7).

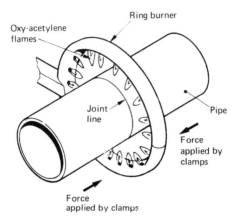

**Fig. 2.7**   Oxy–acetylene pressure welding of small-bore pipes

**Resistance heating**   This relies on the heating effect of a current flowing through a resistance (fig. 2.8). For pressure welding, the components, which are usually round bars, are held in clamps and a high current is passed along the work through the interface. Although the abutting ends of the bars are held in contact under pressure, the oxides at the interface offer a resistance to the passage of the welding current and heat is generated. The rate of heating depends on both the current and the resistance, and the latter is to some extent influenced by the magnitude of the end pressure which brings the faces into close contact, thus increasing the area which can conduct the electricity.

**Induction heating**   This involves the passage of a high-frequency current through a coil surrounding the components to be welded (fig. 2.9). The magnetic field produced in the coil induces eddy currents in the steel workpieces, causing them to heat up.

Hot pressure welding is by no means a universally applicable technique, especially as it is usually restricted to bars, rods, pipes, and narrow strips. It is also difficult to use for 'one-off' work, since a number of trials must be made to ascertain the optimum combination of temperature and pressure.

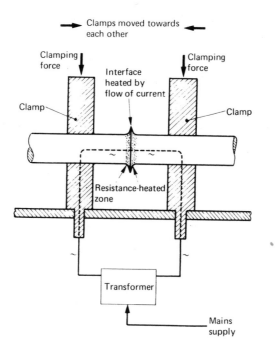

**Fig. 2.8** Principles of resistance butt welding

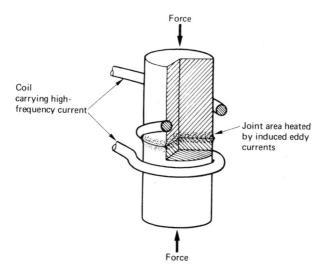

**Fig. 2.9** Using a high-frequency current to heat the interface in pressure welding

15

## 2.5 Friction welding

Without doubt, the most successful development in the field of pressure welding has been that of friction welding. The machine used for this process looks somewhat like a large lathe (fig. 2.10) fitted with two chucks – one driven by a motor, the other fixed. The two parts to be joined are clamped in the chucks and one part is rotated (fig. 2.11). This rotated component must be round in cross-section, but the part held in the fixed chuck can be either a matching section or flat.

**Fig. 2.10** Typical friction-welding machine

When the rotating chuck reaches the welding speed, the parts are brought into contact under a light axial load. As the abutting faces rub together, friction between them generates heat and localised hot plastic zones are produced. With the end load maintained, heat continues to be generated until the whole interface has reached a uniform temperature. At the same time, the plastic metal starts to flow outwards towards the periphery, carrying with it any oxides present at the joint face. When sufficient heating has occurred, the relative rotation of the parts is stopped rapidly and the end load may be increased. The result is a forged pressure butt weld having an excess-metal flash which may be removed by machining (fig. 2.12). Weld times are short, being of the order of 20 to 100 seconds.

The operating parameters for friction welding must be determined by trial and established before production welding begins, so it is clear that the process is not ideally suited to one-off fabrications. Where there is a repetitive element, however, it offers considerable potential and its attraction to the

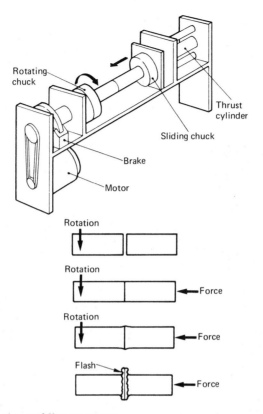

Fig. 2.11  Friction-welding sequence

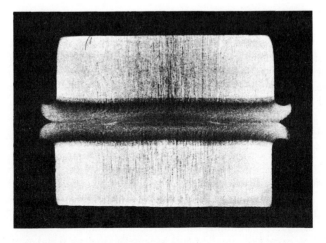

Fig. 2.12  Cross-section of a friction weld

producer of engineering components is increased by the fact that friction-welding units can be readily incorporated into workshops alongside lathes, milling machines, and similar machine tools. A good example of the use of friction welding in a high-volume production line is the manufacture of axle casings for cars and heavy vehicles (fig. 2.13).

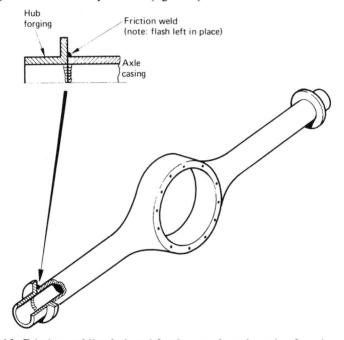

**Fig. 2.13** Friction welding hub-end forgings to the axle casing for a heavy vehicle

The main drawback in applying this process is that at least one of the components being joined must be round in cross-section. The need to rotate one member of the joint can also pose problems – when welding long lengths of pipe, for example. In this case, a solution seems to lie in rotating a wedge-shaped insert between the abutting faces of the pipes, which are being forced towards each other. In this variation, known as *radial friction welding*, heat is generated at two interfaces and plastic deformation occurs as in the conventional process (fig. 2.14).

### 2.6 Principles of fusion welding
All the pressure-welding systems described in sections 2.3 to 2.5 have limitations. In cold pressure welding, large amounts of deformation are required to achieve satisfactory bonding, while in friction welding the machinery for joining any except the smallest components is substantial and expensive. Probably the most critical limitation is that posed by shape and size. In

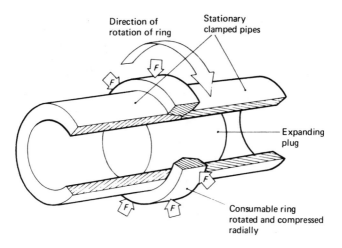

**Fig 2.14 (a)** Joint set-up, motion, and force for radial friction welding

**Fig. 2.14 (b)** Radial friction welding of pipes: completed weld

fabricated structures, the joints which need to be welded more often than not consist of relatively long plate or sheet edges. These are not suitable for pressure welding, mainly because very large forging forces would be required. This would create problems in designing clamps to transmit the force and in accommodating the deformation which would take place. An alternative method of welding must be used for joints of this type. At the end of the last century, engineers, who were trying to devise new production techniques to maintain the impetus of industrial development, discovered that good welds could be achieved by the use of controlled fusion.

19

We can understand the principles of fusion welding by considering the melting of a small area on the surface of a metal plate (fig. 2.15). As the solid metal reaches its melting point, atoms which had been regularly positioned in a lattice structure become free to move about. At the same time, oxide films which existed on the surface of the plate are disrupted and are either dissolved in the molten metal, as with iron or copper, or float on the surface, as with aluminium for example. If we could examine the interface between the solid and liquid metals, it would seem as if the lattice structure within each grain had been cut, leaving the end of each row of solid atoms exposed to molten metal. We would also observe that there are no oxides or other contaminants present – in other words, the solid metal at the interface would be metallurgically clean.

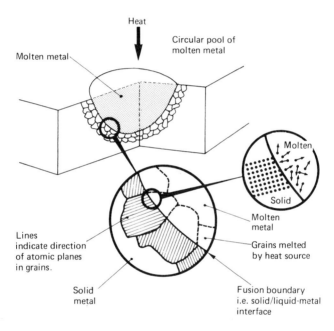

**Fig. 2.15** Formation of molten pool

If the molten pool is allowed to cool, solidification takes place and appears to progress in layers from the solid/liquid-metal interface towards the centre (fig. 2.16). Closer inspection would show that the mechanism of solidification is far more complex. As the temperature of the pool starts to fall, the molten atoms lose their energy. Since the heat is being conducted away from the pool by the mass of the plate, i.e. the heat is flowing through the solid/liquid interface, the atoms in this region are the first to look for permanent sites. The solid atoms at the interface have spare bonds available to which the solidifying atoms can attach themselves, so extending the lattice plane into

20

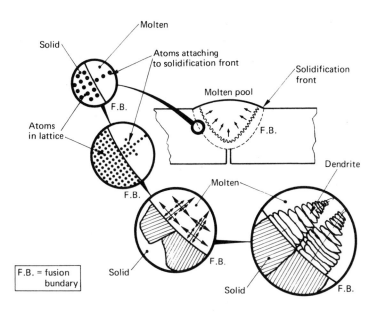

**Fig. 2.16** Stages in solidification and formation of dendrites

the molten pool. Thus a protrusion is produced to which more atoms become joined, in this way extending the lines of solid metal into the weld pool.

Before long, atoms start to attach themselves to the sides of these protrusions, with the result that lateral growth of the solid-metal 'finger' takes place and continues until neighbouring solid regions meet at a boundary. The fingers of solid metal are known as dendrites, and each produces a grain in the solidified weld pool. Once this dendritic pattern has been established, the solid/liquid interface assumes a jagged or sawtooth profile and advances progressively towards the centre of the pool, until all the liquid metal has solidified. The volume which had originally been melted now consists of a collection of dendritic grains bonded to each other at the grain boundaries.

Clearly, the localised bond made by melting just one spot is of limited application, and for most fabrications we need bonding along the complete length of the joint line. This can quite readily be achieved by moving the heat source along the joint line. In this way a series of overlapping pools is in effect produced, and progressive melting and solidification occurs at the leading and trailing edges of the weld pool. Such a weld pool is no longer completely circular but has a round leading edge and an elliptical trailing edge (fig. 2.17). Solidification occurs at the rear of the pool only when the heat source has moved forward so that the flow of heat into the plate reduces the temperature of the trailing edge to the solidification point of the metal. The mechanics of grain growth are the same as those described for the stationary circular pool, except that the dendrites point upwards and forwards in the direction of the centre line (fig. 2.18).

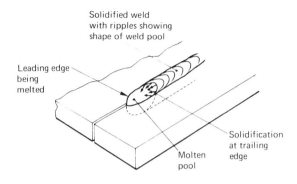

Solidified weld
with ripples showing
shape of weld pool

Leading edge
being
melted

Solidification
at trailing
edge

Molten
pool

Fig. 2.17 Progressive melting and solidification in a welded joint

Direction of travel

F.B.

F.B.

Fig. 2.18 Etched section of weld, showing grains pointing in the direction of travel

## 2.7 Heat sources for fusion welding

To make a fused joint in the way described above requires a heat source capable of creating localised fusion in a controlled manner. Such a heat source must have certain basic attributes if it is to be useful in practical situations:

a) It must operate at a temperature significantly higher than the melting point of the metal being welded. If the temperature differential is small, the heat flows away from the joint almost as quickly as it is being supplied, with the result that it is difficult to raise the temperature to the melting point and at the same time keep the width of the melted zone reasonably small.

22

b) It follows that the heat should be concentrated into a small area if the weld pool is to be constrained or restricted to predetermined proportions. Sources which may spread their heat over a large area, as in brazing and soldering, cannot normally be used for welding.

c) There must be an adequate heating capacity. The total amount of heat, or the rate at which heat is required, depends not only on the physical properties of the metal but also on the joint configuration and dimensions.

d) The heat source itself must be capable of regulation, so that conditions can be set to suit the joint. These conditions must remain constant throughout the welding operation.

The most common heat sources used in current industrial welding practice are an oxygen–fuel-gas flame, an electric arc, heat generated by resistance at an interface, and heat generated by resistance in a slag bath.

**An oxygen–fuel-gas flame** (fig. 2.19)  By burning a mixture of oxygen and a fuel gas at the outlet of an orifice in a tube or nozzle, it is possible to achieve quite high temperatures. Unfortunately, with most oxygen–fuel-gas mixtures these temperatures are too low for the welding of anything other than a few low-melting-point metals such as lead, zinc, and tin. The one exception is provided by acetylene. When mixed with oxygen in the correct proportions, this gas burns with a flame temperature of about $3100^{\circ}$C, which is adequate for many welding applications.

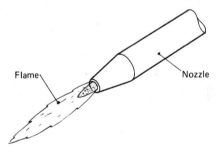

**Fig. 2.19**  Oxygen–fuel-gas flame

**An electric arc** (fig. 2.20)  A low-voltage (20 to 40 V), high-current (30 to 1000 A) electric arc between the end of a rod electrode and the flat surface of a plate constitutes a very effective heat source which forms the basis of the majority of the techniques used to weld plates, pipes, and sections.

**Resistance heating at an interface** (fig. 2.21)  When discussing pressure welding, we saw that a butt joint between two pieces of rod can be heated, to make the abutting ends plastic, by passing a high current through the interface. In that operation there was no intention of reaching the melting point of the rod material, but, if the current had been allowed to flow for a longer time or if the current density had been increased, the faces might well

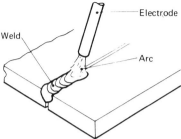

**Fig. 2.20** Electric arc

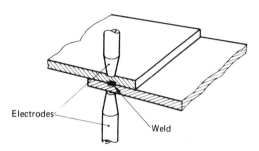

**Fig. 2.21** Resistance heating at an interface

have melted. This principle can be applied to the production of a fused spot at the interface of two overlapping sheets. If, after melting has occurred, the spot is allowed to solidify, there will be a continuous metallic bridge over a limited area of the interface, i.e. a spot weld.

**Resistance heating in a slag bath** (fig. 2.22)    An alternative way of using resistance heating is to pass a high current through a bath of molten slag. Mineral or ceramic slags are usually good insulators when cold, but they conduct electricity when they are melted. Their electrical conductivity is low, which means that they offer a resistance to the flow of current, and there is therefore considerable heat generated in the body of the molten slag. This heat can be utilised to melt metal plates in contact with the slag bath. Although widely used in metallurgical extraction processes, slag-bath heating features in only one major welding process: electro-slag welding and, its variant, consumable-guide welding.

## 2.8 Protection of the weld pool
All the heat sources discussed above produce a fused region having a temperature well in excess of the melting point of the metal being welded. If the weld pool, once established, is exposed to open atmosphere, it absorbs oxygen and/or nitrogen and/or hydrogen, depending on the composition of the parent metal.

24

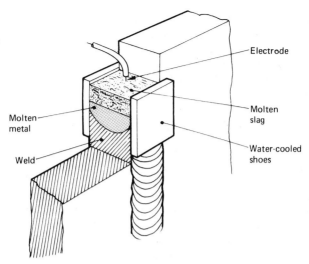

**Fig. 2.22** Resistance heating in a slag bath

Small quantities of gas either dissolved in or combined with weld metal can usually be tolerated without detriment to the properties of the joint. In general, however, the presence of significant amounts of dissolved or combined gases is undesirable, since it often results in the formation of gas pores or voids in the completed weld. In addition, the properties or integrity of the joint can be impaired. For example, the ductility of aluminium weld metal is affected by the existence of oxide films at the grain boundaries; the electrical conductivity of copper is reduced by an increase in the oxygen content; dissolved hydrogen in high-strength-steel weld metal can lead to cracks in the joint.

Clearly, any viable welding system must include some method of preventing atmospheric contamination. In current practice two basic techniques are employed. From the point of view of historical development, the first of these is to blanket the weld pool with molten flux, forming a slag layer which is impervious to the passage of gases. The simpler alternative is to replace the air in the vicinity of the heat source and weld pool by a gas which does not react with the molten metal and is therefore in this sense inert.

### 2.9 Typical fusion-welding processes
It can now be seen that for each fusion-welding process we must consider two aspects: firstly the provision of a suitable source of heat and secondly the need to protect the weld pool against atmospheric contamination. In the following pages, profiles of some of the commonly used fusion-welding processes are given. The collection is not intended to be exhaustive; the aim is only to illustrate those systems which have made a major contribution to the success of modern fabrication engineering. Other, less widely used processes could be analysed and presented in the same way.

25

*Oxy–acetylene welding* (figs 2.23 to 2.25)

*Alternative name*    Gas welding

*Type of operation*    Manual

*Heat source*    Fuel-gas flame

*Shielding*    Products of combustion; flux for metals other than steel

*Heat input*    See chapter 3

*Mode of operation*    An oxygen and acetylene mixture is burnt at the tip
of a specially designed nozzle which is fitted to a torch body. The welder
uses the flame to melt the parent metal to form a weld pool. Filler metal,
if required, is added separately by manual feeding of a wire into the leading
edge of the weld pool. The welder moves the torch to achieve uniform
progressive fusion.

*Typical applications*    Light fabrications such as ventilation ducts; small-bore
pipework for heating and chemical systems; motor-vehicle repair

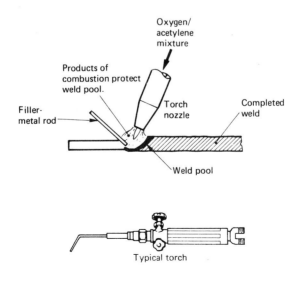

**Fig. 2.23**  Schematic diagram of oxy–acetylene welding

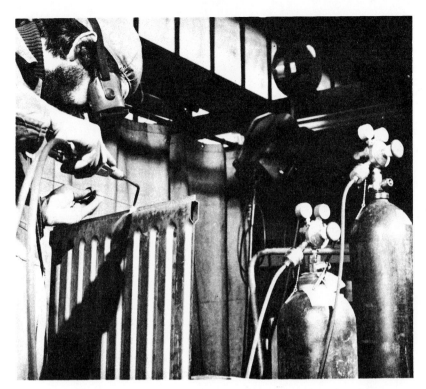

Fig. 2.24 General view of oxy–acetylene welding

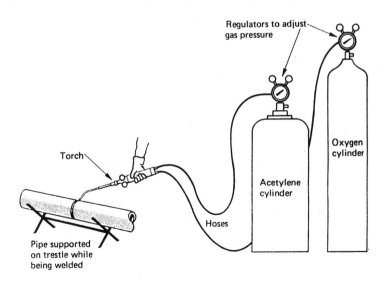

Regulators to adjust gas pressure

Oxygen cylinder

Torch

Acetylene cylinder

Hoses

Pipe supported on trestle while being welded

Fig. 2.25 Equipment for oxy–acetylene welding

*Tungsten arc gas-shielded (TAGS) welding* (figs 2.26 to 2.28)
*Alternative names*    Tungsten inert-gas (TIG) welding; gas tungsten arc
welding (GTAW); Argonarc welding (BOC Ltd trade name)
*Type of operation*    Manual
*Heat source*    Arc
*Shielding*    Inert gas
*Current range*    10 to 300 A
*Heat input*    0.2 to 8 kJ/s
*Mode of operation*    An arc is established between the end of a tungsten
electrode and the parent metal at the joint line. The electrode is not melted
and the welder keeps the arc gap constant. The current is controlled by the
power-supply unit. Filler metal, usually available in 1 m lengths of wire, is
added to the leading edge of the pool as required. The molten pool is
shielded by an inert gas which replaces the air in the arc area. Argon is the
most commonly used shielding gas.
*Typical applications*    High-quality welds in metals such as aluminium,
stainless steels, Nimonic alloys, and copper in chemical plant; sheet work
in aircraft engines and structures

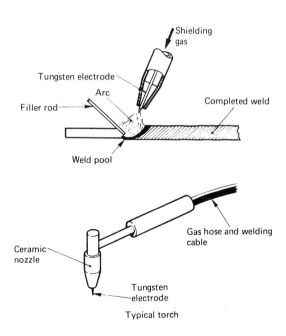

**Fig. 2.26**  Schematic diagram of TAGS welding

28

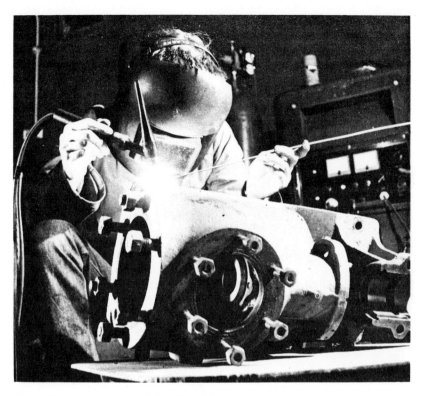

**Fig. 2.27** General view of TAGS welding

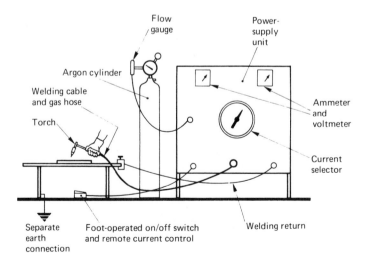

**Fig. 2.28** Equipment for TAGS welding

29

*Manual metal arc (MMA) welding* (figs 2.29 to 2.31)
*Alternative names*    Stick-electrode welding; electric arc welding; shielded metal arc welding (in USA)
*Type of operation*    Manual
*Heat source*    Arc
*Shielding*    Principally flux; some gas generated by flux
*Current range*    25 to 350 A
*Heat input*    0.5 to 11 kJ/s
*Mode of operation*    The welder establishes an arc between the end of the electrode and the parent metal at the joint line. The arc melts the parent metal and the electrode to form a weld pool which is protected by the molten flux layer and gas generated by the flux covering of the electrode. The welder moves the electrode towards the weld pool to keep the arc gap at a constant length. The current is controlled by the power-supply unit. Electrodes are normally 460 mm long. When the electrode has been melted to a length of about 50 mm, the arc is extinguished. The solidified slag or flux is removed from the surface and the weld is continued with a fresh electrode.
*Typical applications*    Fabrication of pressure vessels, ships, structural steelwork; joints in pipework and pipelines; construction and repair of machine plant

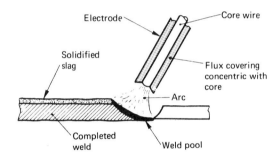

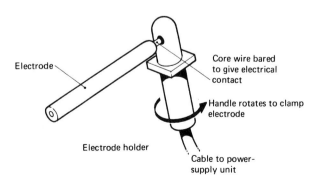

**Fig. 2.29**  Schematic diagram of MMA welding

Fig. 2.30  General view of MMA welding

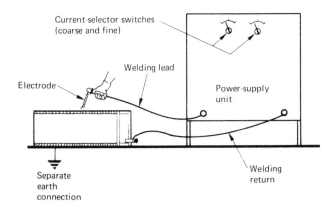

Fig. 2.31  Equipment for MMA welding

*Metal arc gas-shielded (MAGS) welding* (figs 2.32 to 2.34)

*Alternative names*   Metal inert-gas (MIG) welding; metal active-gas (MAG) welding; gas metal arc welding (GMAW); semi-automatic welding; $CO_2$ welding

*Type of operation*   Manual, but can be used with a mechanised traversing system

*Heat source*   Arc

*Shielding*   Gas, which must not react with the metal being welded

*Current range*   60 to 500 A

*Heat input*   1 to 25 kJ/s

*Mode of operation*   An arc is established between the end of the electrode and the parent metal at the joint line. The electrode is fed at a constant speed by a governed motor. The electrode feed rate determines the current. The arc length is controlled by the power-supply unit, and the welder is required to keep the nozzle at a fixed height above the weld pool (usually about 20 mm). The arc area and the weld metal are protected by a gas which is chosen to suit the metal being welded. Gases commonly used are argon, argon mixed with 5% oxygen or 20% carbon dioxide, and pure carbon dioxide.

*Typical applications*   Medium-gauge fabrications such as earth-moving equipment, plate, and box girders; sheet-metal work for car bodies

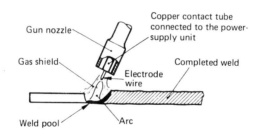

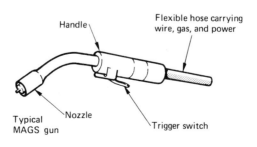

**Fig. 2.32**  Schematic diagram of MAGS welding

**Fig. 2.33** General view of MAGS welding

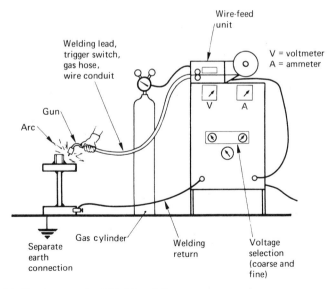

**Fig. 2.34** Equipment for MAGS welding

*Submerged-arc welding* (figs 2.35 to 2.37)
*Type of operation*    Mechanised
*Heat source*    Arc
*Shielding*    Granular flux
*Current range*    350 to 2000 A
*Heat input*    9 to 80 kJ/s
*Mode of operation*    An arc is maintained between the end of a bare wire electrode and the parent metal. The current is controlled by the power-supply unit. As the electrode is melted, it is fed into the arc by a servo-controlled motor. This matches the electrode feed rate to the speed at which the electrode is melting, thus keeping the arc length constant. The electrode and drive assembly is moved along the joint line by a mechanised traverse system.

The arc operates under a layer of granular flux (hence 'submerged' arc). Some of the flux melts to provide a protective blanket over the weld pool. Unmelted flux is recovered and re-used.
*Typical applications*    Joints in thick plate in pressure vessels, bridges, ships, structural work, welded pipe

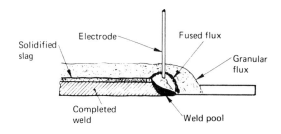

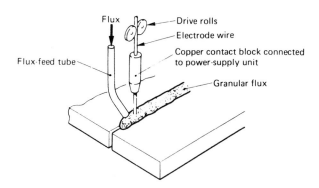

**Fig. 2.35**  Schematic diagram of submerged-arc welding

**Fig. 2.36** General view of submerged-arc welding

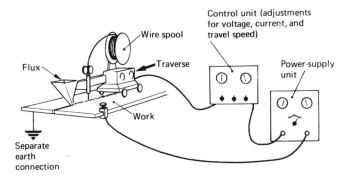

**Fig. 2.37** Equipment for submerged-arc welding (note that other types of traverse mechanism can be used)

*Resistance spot welding* (figs 2.38 to 2.40)

*Type of operation*    Automatic

*Heat source*    Resistance heating at an interface

*Shielding*    None required

*Current range*    100 to 50 000 A

*Heat input*    See chapter 11

*Mode of operation*    The work, which is usually in the form of a lap joint, is gripped between two copper electrodes. A high current at a low voltage flows through the parent metal between the electrodes. At the interface, heat is generated by the resistance offered to the current flow. A spot or slug of metal is melted and bridges the interface. The current flows for only a short time (typically 0.06 to 3 seconds). When the current is switched off (automatically), the weld solidifies under pressure.

*Typical applications*    Light fabrications from pressed sheet, such as car bodies and domestic washing machines. Also used for high-quality work in aircraft engines.

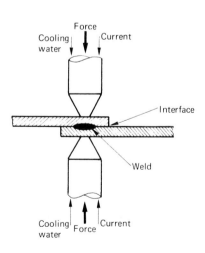

**Fig. 2.38** Schematic diagram of resistance spot welding

**Fig. 2.39** General view of resistance spot welding

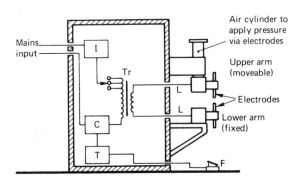

**Fig. 2.40** Principal components of a resistance spot welder
C – on/off contactor   T – timer   F – foot switch to initiate weld cycle
Tr – transformer   L – flexible leads connecting transformer to electrodes
I – ignitrons to control current (used only on larger machines)

# 3 Essential parameters in fusion welding

## 3.1 Parent-metal fusion

We will now consider fusion welding in more detail, beginning with the factors which influence the way the heat source melts the parent metal, since control of the weld pool is the key to making a weld of good quality. Our interpretation of what is meant by the expression 'quality of the finished weld' is very closely related to the type of service for which the joint is required, and we will be discussing this aspect at length in chapter 7; for the moment it is sufficient to emphasise that for a weld to be acceptable there must be good bonding between the weld metal and the plate material. This can occur only when the surface of the parent metal has been melted before the added electrode or filler metal is allowed to flow into the joint. If this sequence is not followed, there will be a lack of fusion at the boundary of the weld metal, since the molten filler will be chilled by the unmelted parent metal and will freeze without having provided enough heat to melt the joint surfaces. An examination of an etched macrosection of a weld made in this way (fig. 3.1) shows up the lack of fusion, and it is clear that the resultant joint would be weak in transverse tension.

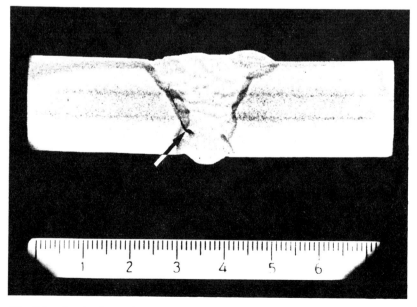

Fig. 3.1   Lack of side fusion in a butt weld (scale numbered in cm)

How, then, can we avoid this situation in practice? In the case of oxy–acetylene and TAGS welding, the problem is relatively straightforward since the filler metal is added separately. The first task, therefore, is to ensure that the edges of the joint have been melted and blended into the weld pool before the filler-metal additions are made. This melting must be achieved as quickly as possible, and its extent must be limited if the welder is to keep control of the weld pool. The filler metal is then added to the leading edge of the pool while the heat source is used to supply sufficient heat to the joint to prevent premature solidification.

With the manual consumable-electrode processes – i.e. manual metal arc and MAGS welding – molten metal is being transferred across the arc throughout the whole of the welding operation, and the welder must manipulate or direct the arc so that fusion of the joint faces is obtained. This is helped by the use of a force which acts along the centre line of the arc to control the flow of metal in the weld pool. In practice, therefore, the welder pays attention to maintaining the correct angle between the electrode and the surface of the weld, since this influences both the distribution of the heat from the arc and the flow of the weld metal over the fused faces.

## 3.2 Heat input

The success of the welding operation also depends on the heat input to the joint. As soon as heat is supplied to the area to be welded, either by an arc or by a fuel-gas flame, it will start to flow away into the metal on either side, since this is at a lower temperature and a temperature gradient is established. This means that, if we are to achieve melting, the rate at which heat is being supplied to the joint must be greater than the rate at which it is flowing into the parent metal.

It follows that the thermal conductivity of the parent plate is probably one of the most important considerations when choosing welding conditions. The various metals used in fabrication conduct heat at different rates. Specific values can be attributed to the thermal conductivities of typical metals, but it is easier to compare them by using an arbitrary scale. Pure copper has a very high thermal conductivity and, by giving it a value of 100, Table 3.1 can be drawn up.

**Table 3.1** Relative thermal conductivities

| Metal | Relative thermal conductivity |
|-----------|:---:|
| Copper | 100 |
| Aluminium | 62 |
| Steel | 14 |
| Lead | 8 |

During a welding operation, heat will be conducted away from the area much faster with copper than with steel. This means that, if we are to achieve fusion, heat must be supplied to a copper joint at a greater rate compared with steel of a similar thickness. On some occasions this may not be practicable, since it may call for a heat input which is in excess of that which can be provided by the process we are using. This situation is affected by the melting point of the metal. Although aluminium has a high thermal conductivity, it is not too difficult to produce fusion as the melting point is only about $600^\circ$C, compared with $1053^\circ$C for copper, but it may be difficult to restrict the width of the melted zone.

Heating the parent metal before welding – i.e. preheating – can assist in reaching the melting point rapidly, by reducing the temperature difference between the weld and the plate. Preheating in this context has a dual role, since the thermal conductivity of a metal falls with increasing temperature. Thus by heating the plate the thermal conductivity is lowered, and this in turn reduces the rate of flow of heat.

Finally, we must consider the importance of the cross-sectional area of the conductor. Normally in welding we are concerned more with thickness than with cross-sectional area as such, and, in general, the thicker the component which is being welded the faster the heat is conducted away from the joint line. We must also recognise that only rarely do we weld just one piece of metal. Usually the joint contains two or more members, each providing a path along which heat can flow; for example, a 'T' joint has three possible heat paths – the vertical leg and both directions in the horizontal member – and so cools faster than a butt joint, which has two.

To summarise, the parameters involved in effective melting of the parent metal during welding are

a) metal thickness, and joint type,
b) thermal conductivity,
c) temperature of the parent metal before welding,
d) melting point,
e) electrode angles and manipulation,
f) heat input.


## 3.3 Measuring heat input in arc welding

When considering heating effects, it is usual to express the amount of heat energy in joules. In arc welding it is difficult to measure heat input directly in joules, but we can deduce it from the power input.

Power is the rate at which energy is converted from one form into another. When dealing with a welding arc, almost all the electrical energy is converted to heat – only a small proportion is used in the generation of the bright light and ultra-violet radiation given off by the arc. We can determine the power input, in watts, as we can readily measure current and voltage:

arc power input (watts) = arc voltage x arc current

For example, with an MMA electrode operating at 150 A, 25 V,

power input = 25 V x 150 A = 3750 watts = 3.75 kW

This applies to both d.c. and a.c. welding, though for the latter r.m.s. (root-mean-square) voltage is used in the calculation.

From the relationship between power and energy,

1 watt = 1 joule per second

hence, in the example above, the energy input to the arc is 3750 joules/second or 3.75 kJ/s.

We must remember, however, that this is the energy input to the arc. In other words, it tells us that, as long as the welder is using the arc, heat is being generated at a rate of 3.75 kJ/s. If the arc was kept stationary this would be the heat input to the weld at one point, but in practice the welder moves along the joint line. The heat is therefore distributed, and the input at any given point depends on the travel speed. Knowing the energy being supplied to the arc and the travel speed, it is possible to calculate the heat input and to express it as the amount of heat per unit length of weld run (i.e. in joules/mm).

Suppose a welder is using the TAGS process to butt weld some stainless-steel sheet which has a thickness of 1.6 mm. The current would be about 60 A and with a 2 mm arc length the arc voltage in argon shielding gas would be 14 V. The travel speed would depend on the actual component, but let us take 120 mm/min as typical. The arc power would be

60 A x 14 V = 840 watts

= 840 joules/second

The welder moves the torch along the joint at the speed of 120 mm/min; hence

in 1 minute the distance moved = 120 mm

in 1 second the distance moved = $\frac{120}{60}$ mm = 2 mm

The heat input is therefore 840 joules every 2 mm, i.e. 420 J/mm.

In general terms,

$$\text{heat input (joules/mm)} = \frac{\text{current (A) x arc voltage (V) x 60}}{\text{travel speed (mm/min)}}$$

or, using kilojoules,

$$\text{heat input (kJ/mm)} = \frac{\text{current (A) x arc voltage (V) x 60}}{\text{travel speed (mm/min) x 1000}}$$

### 3.4 Heat input in oxy–acetylene welding
By contrast with arc welding, in which electrical energy is converted to heat at the surface of the pool, oxy–acetylene welding depends on a chemical reaction to generate heat which is then transferred to the work.

The oxy–acetylene flame used for welding consists of two zones: an inner cone which is well defined and an outer envelope which is more diffuse. These two zones correspond to the two stages of combustion which occur in the flame (fig. 3.2). As the mixed acetylene and oxygen emerge from the

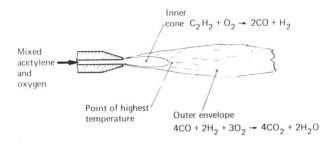

**Fig. 3.2** Simplified structure of an oxy–acetylene flame

nozzle, the acetylene (chemical formula $C_2H_2$) burns to give carbon monoxide and hydrogen:

$$C_2H_2 + O_2 \longrightarrow 2CO + H_2 + \text{heat}$$

Carbon monoxide and hydrogen are both combustible gases and they in turn burn in the outer envelope when they are mixed with oxygen from the surrounding air:

$$4CO + 2H_2 + 3O_2 \longrightarrow 4CO_2 + 2H_2O + \text{heat}$$

The total quantity of heat produced by these reactions depends on the amount of acetylene which is burnt. If we require more heat, the flow rate of acetylene is increased and the oxygen supply is adjusted to give the correct type of flame. This involves a change in the size of the orifice through which the mixed gases flow, and the nozzles are graded accordingly.

The total amount of heat available is about 55 kJ per litre of acetylene. Unfortunately, not all of this heat can be used directly to produce fusion of the joint faces. The highest temperature in the oxy–acetylene welding flame is at the tip of the inner cone (fig. 3.2), and, for the most efficient operation of the system, the flame should be positioned so that this point is just above the surface of the weld pool. At this point in the flame, the only heat available for melting is that developed by the first reaction:

$$C_2H_2 + O_2 \longrightarrow 2CO + H_2$$

This represents approximately 35% of the total heat produced, and for each litre of acetylene burnt only about 19 kJ of the heat produced is used to fuse the parent metal. The heat from the second-stage reaction is not wasted, however, since the outer envelope of the flame covers the surface of the parent metal around the weld and effectively preheats the area.

Although the temperature of the acetylene flame is high enough for welding, and sufficient heat can be supplied, the rates of travel are slower than with arc welding. Briefly, there are two reasons for this. Firstly, the rate at which heat is transferred from the heat source to the work depends on the difference between the melting point of the metal and the temperature of the source. In oxy–acetylene welding the differential is not very large, and

42

heat transfer is relatively slow. This is why it is most important to position the inner cone correctly — if there is a large gap between the tip of the cone and the surface of the weld pool, heat is transferred from a region of lower temperature and the process will therefore be less efficient. Secondly, in arc welding much of the heat used for melting is generated at the surface of the parent metal itself and hence does not need to be transferred.

### 3.5  Summary of the main factors governing heat input
When summarising the factors governing heat input, we must therefore distinguish between arc welding and gas welding.

In arc welding, the main parameters are

arc current,
arc voltage,
travel speed.

In oxygen–fuel-gas welding, the main parameters are

fuel-gas flow rate,
correct positioning of flame,
travel speed.

### 3.6  Penetration and dilution in arc welding
So far, little has been said about the way in which the parent metal is melted in arc welding. In essence, an arc is the passage of electricity across a gap between two electrodes. For this to be possible, the atmosphere in the gap must be ionised – this means that the atoms of gas have lost a free electron and are positively charged. The electrons travel from the cathode (negative) to the anode (positive), giving the flow of current. Some electrical energy is converted to heat in the arc column (fig. 3.3) and is radiated into the surrounding air, but the majority is released at the end of the electrode and at the plate surface.

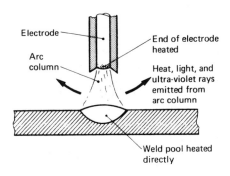

**Fig. 3.3**    Distribution of heat in a welding arc

43

The proportion of the total energy supplied to the arc which is used to melt the plate will depend on whether the parent metal is an anode or a cathode, i.e. whether it is connected to the positive or negative pole of the power-supply unit. The choice of polarity is governed by a number of considerations such as arc stability, rate of electrode melting, transfer of melted metal from the electrode to the weld pool, and removal of oxides from the surface of the parent metal. These vary from one process to another, and it is not possible to generalise.

The depth to which the parent metal has been fused is usually referred to as the *penetration*. This may be seen simply as melting into the plate, as in the case of fillet-welded 'T' joints, or as the extent to which the abutting edges in a joint have melted (fig. 3.4).

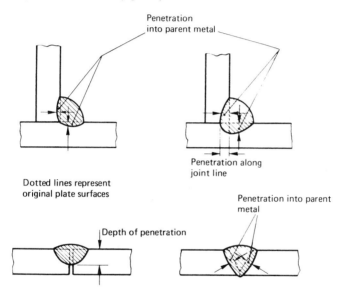

**Fig. 3.4**     Penetration in 'T' and butt joints

Since both voltage and current contribute to heating in the arc, a change in either will have an effect on the penetration. During welding, however, the magnitude of the voltage is usually decided by other considerations. The width of the weld and the surface profile are dependent on the arc length, which in turn is closely related to voltage. In general terms, high voltages give long arc lengths, wide flat welds, and the risk of arc instability. At the same time, oxygen and nitrogen may be drawn into the arc column. On the other hand, very low voltages (i.e. short arcs) can result in narrow welds having a high profile with the possibility of lack of fusion at the edge of the weld. The welder, therefore, choses a voltage and arc length which give a stable arc and a satisfactory weld surface profile. Once the voltage has been fixed, current becomes the main factor in the control of penetration.

An important aspect of penetration in arc welding is the effect it has on weld-metal composition. The molten pool is made up of a mixture of electrode and parent metal. The proportions will be determined to a large extent by the heat input, since this controls both the rate of electrode melting and the penetration. It is useful to view the weld metal as metal deposited from the filler wire or electrode diluted by mixing with the melted material from the workpiece. By convention, dilution is expressed as the percentage of melted parent metal in the weld (fig. 3.5) and can be used to predict individual alloy content in the molten pool.

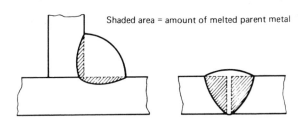

Shaded area = amount of melted parent metal

**Fig. 3.5**   Effect of dilution on composition

$$\text{Dilution} = \frac{\text{shaded area}}{\text{total area of weld}} \times 100\%$$

$$c_w = c_f + \frac{D}{100}(c_p - c_f)$$

where
$c_w$ = percentage of a given element in the weld
$c_f$ = percentage of this element in the filler or electrode
$c_p$ = percentage of this element in the parent metal
$D$ = percentage dilution

This concept is of particular value in designing procedures to avoid cracking when fabricating some stainless steels and aluminium alloys – the composition must be carefully controlled to ensure that the weld metal has adequate strength at solidification. Again, when coating the edge of a carbon-steel blade with wear-resistant weld metal, the properties of the deposited layer may be impaired if too much steel is melted into it. In such cases, if the acceptable dilution can be determined, it is possible to specify the maximum allowable penetration.

### 3.7 Penetration and edge preparation
The thickness of metal which can be butt-welded using square edges for the joint depends on the amount of penetration available. In arc welding, the limiting factor is current. With the MMA and TAGS processes the maximum

thickness is 3 mm if the weld is made from only one side or 5 mm if a second run can be deposited on the reverse side (fig. 3.6). The same limitations apply to MAGS welding at currents below 200 A; but with high currents, say 400 to 450 A, plates having a thickness of 6 mm can be butt-welded from one side with square edges.

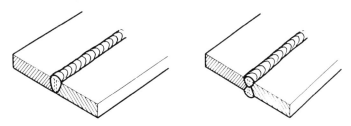

**Fig. 3.6**    Square-edge butt-welded joints

With thicker material, the edges must be cut back to provide access for the arc. The simplest edge preparation (fig. 3.7) is a single bevel. When the joint is assembled, a groove is formed which is filled by depositing a number of weld runs into it. Each run is fused into both the surface of the preceding weld and the side walls of the groove, to ensure bonding through the complete thickness of the joint. The first weld or pass to be deposited is known as the root run (fig. 3.8). It often requires considerable skill to produce a satisfactory root run, since the welder must fuse the root faces but at the same time keep control of the width of the weld bead as the molten metal is unsupported on the underside. The risk of producing either excessive penetration or lack of fusion in a root run is always very great, and in some areas of work it is common practice to remove the root run by gouging a groove along the line of penetration after the 'V' has been filled. The joint is then completed by depositing a sealing run from the reverse side.

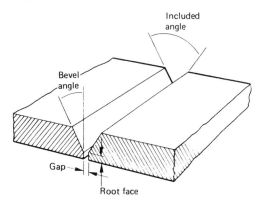

**Fig. 3.7**    Single-V preparation

46

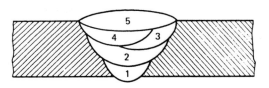

| | | |
|---|---|---|
| 1 | = | root run |
| 2,3,4 | = | filling runs |
| 5 | = | final or capping run |

Fig. 3.8    Typical welding sequence for a single-V butt weld

The root gap plays an important part in controlling penetration in the root of a joint. The size of the gap depends on the process and the application, i.e. plate or pipe, type of joint, and position of welding. It is most important that the gap is constant – if it varies along the joint, the welder will have great difficulty in achieving consistent penetration. Where it is not possible to ensure uniformity of fit-up or adequately to control root

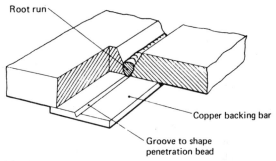

(a)    Temporary backing bar, with groove, to
support root run (removed after welding)

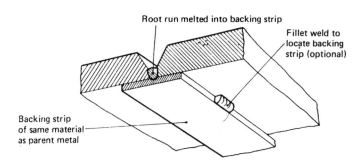

(b)    Permanent backing strip

Fig. 3.9    Temporary and permanent backing for butt welds

47

fusion, some form of backing can be used. This may be temporary, in that it is removed after welding, or the root run may penetrate into a strip of metal which becomes a permanent part of the joint (fig. 3.9).

Various types of edge preparation can be used (fig. 3.10), and the choice of the most suitable is influenced by a number of factors. Some of these, not necessarily in order of importance, are

a) type of process,
b) type of work,
c) position of welding (see fig. 5.3)
d) access for arc and electrode,
e) volume of deposited weld metal,
f) dilution,
g) cost of preparing edges,
h) shrinkage and distortion.

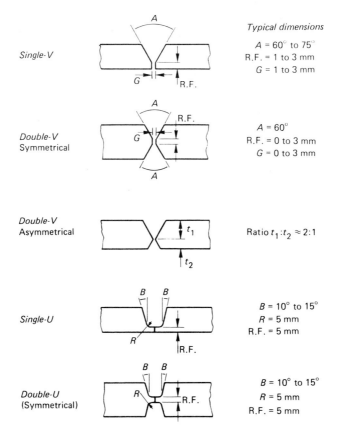

Fig. 3.10 Typical edge preparations for butt joints

48

# 4 Control of welding parameters

## 4.1 The importance of control

In chapter 3 we developed the idea that for a weld to be considered of acceptable quality there must be bonding between the weld metal and parent material, but it would be pointless if this occurred only at intervals along the joint. Clearly, if the weld is to have full strength, there must be bonding at all points throughout the length of the joint, which means that the heat input must be controlled to ensure it is always above the minimum level required for the thickness being welded. It is then left to the welder to use the correct manipulative technique and travel speed to achieve continuity.

Having achieved fusion, the welder must also aim to keep the size of the weld constant within certain limits and to produce a uniform profile or shape. Consistency in this sense is important both to the designer, who will specify a weld size to suit the load which the joint is to carry, and to the estimator, who will want to know that the time to deposit a weld will remain reasonably reproducible.

These considerations pose the question: how do we control the heat input? The answer to this obviously depends on the nature of the heat source, and for the two types which we have considered so far we need to analyse the main parameters governing input.

## 4.2 Control in oxy–acetylene welding

In the case of oxy–acetylene welding, the analysis is relatively straightforward. The heat used to produce melting is provided by the combustion of acetylene at the orifice of the nozzle. The more acetylene supplied, the greater will be the heat generated, which means that the flow rate of acetylene must be controllable. When the oxy–acetylene flame is used for welding, the heat input to the joint will also depend on the efficiency of the combustion. The maximum total heat is produced when the acetylene is completely burnt in an oxidising flame, i.e. a flame containing an excess of oxygen over that needed to combine with the acetylene, but this setting is rarely employed for welding since it does not give the hottest flame and can cause oxidation of the weld metal. More normally a ratio of acetylene to oxygen is chosen which produces a neutral flame – i.e. one in which there is no excess of either gas. The relative amounts of acetylene and oxygen are adjusted by means of flow valves built into the torch. The gas supplied to the nozzle is therefore a controlled mixture of oxygen and acetylene which must be measured and regulated.

The use of this principle can be seen in a simplified layout of a typical oxy–acetylene system (fig. 4.1). The regulators not only reduce the pressure

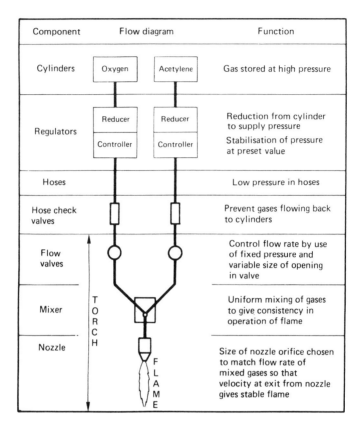

| Component | Flow diagram | Function |
|---|---|---|
| Cylinders | Oxygen / Acetylene | Gas stored at high pressure |
| Regulators | Reducer / Reducer / Controller / Controller | Reduction from cylinder to supply pressure. Stabilisation of pressure at preset value |
| Hoses | | Low pressure in hoses |
| Hose check valves | | Prevent gases flowing back to cylinders |
| Flow valves | | Control flow rate by use of fixed pressure and variable size of opening in valve |
| Mixer | | Uniform mixing of gases to give consistency in operation of flame |
| Nozzle | | Size of nozzle orifice chosen to match flow rate of mixed gases so that velocity at exit from nozzle gives stable flame |

**Fig. 4.1**  Schematic layout of oxy–acetylene system

of the gas from that of the cylinders to the level required to achieve correct flow rate and flame stability, but also keep the reduced pressure constant, thus ensuring uniformity of operation. The flow valves in the torch alter the amounts of acetylene and oxygen being supplied to the mixer. These must be adjusted in conjunction with the regulators, since the rate of flow depends on both the pressure at the entry to the valve and the size of the valve opening. The mixed gases are fed to the nozzle, where the diameter of the orifice must be chosen to match the volume of the acetylene and oxygen mixture being used, so that the velocity of the gas issuing from the nozzle enables a stable flame to be maintained. A sectional view of a modern oxy–acetylene welding torch (fig. 4.2) shows how these features are incorporated.

## 4.3  Control in arc welding

Turning now to the use of an arc for welding, we can first summarise its main features:

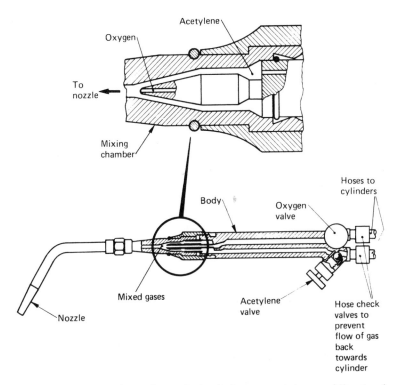

**Fig. 4.2** Diagrammatic sections of a typical oxy–acetylene welding torch

a) A welding arc is the passage of electricity across a gap between an electrode and the parent metal or work.
b) Heat is generated at the surfaces of the electrode and of the parent metal.
c) Heating at the electrode can cause melting and the transfer of metal to the weld pool.
d) Heating at the parent metal produces fusion of the joint faces.
e) Heat input to the weld is a function of arc voltage, arc current, and travel speed.
f) Arc length is related to arc voltage.

*The arc characteristic*
As with the oxy–acetylene flame, in arc welding the welder adjusts the travel speed to give a uniform weld, and the requirement is for an arc with a constant voltage and current. When the arc is operating in a stable manner, the voltage and current are related. The relationships can be shown graphically by plotting the appropriate value of arc voltage for various currents (fig. 4.3). This graph is known as an *arc characteristic*.

It will be seen that the arc does not behave like a simple resistance, for which Ohm's law predicts that current increases proportionally with voltage.

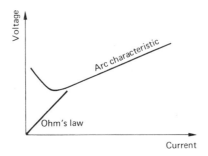

**Fig. 4.3** Typical arc characteristic compared with Ohm's law

The graph in fig. 4.3 also shows the Ohm's law characteristic for a resistor of 1.0 ohms. By comparison, the arc voltage varies only slightly over a wide range of currents and the curve does not pass through the origin. The slope of the arc characteristic depends on the metals involved, the atmosphere in the arc gap, and the arc length.

### 4.4 Arc-length control in TAGS welding

Typical characteristics for four arc lengths between a tungsten electrode and copper in an argon atmosphere are shown in fig. 4.4. From these we can read off the values of voltage at various currents for different arc lengths and plot the results (fig. 4.5) to show that, for a particular current, the voltage increases as the arc is lengthened. This observation is of considerable importance in solving the practical problem of controlling the arc.

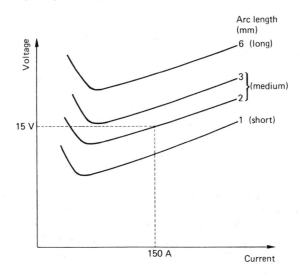

**Fig. 4.4** Arc characteristics for welding copper in an argon atmosphere (TAGS welding)

52

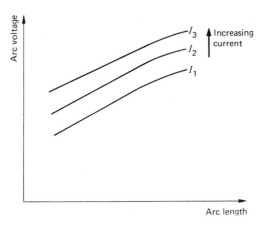

**Fig. 4.5**  Relationship between arc voltage and arc length in TAGS welding

For a start, let us take the manual welding of copper sheet using the TAGS process and suppose that the operation requires 150 A with an arc length of 2 mm. The power-supply unit is set to give an output of 150 A. The operator strikes the arc, establishes a weld pool, and travels along the joint. The arc characteristics in fig. 4.4 tell us that at 150 A for a 2 mm arc to be operating satisfactorily the voltage should be 15 V. This value will be maintained as long as the power-supply unit continues to deliver 150 A and the welder keeps the arc length at 2 mm.

In reality many things may cause the arc length to vary, with the result that the voltage will rise or fall in sympathy and the operating point will move from one characteristic to another. If the power supply was able to provide only one voltage at the selected current, the arc would be forced to work at an incorrect voltage and it would become difficult for the operator to control the melting in the joint, i.e. the arc would be unstable. For stable operation the power-supply unit must allow the voltage to vary while keeping the current substantially constant. It follows from this that the output of the power-supply unit must be designed to meet the specific requirements of the TAGS welding process. It is usual to describe the output in terms of the voltage and current relationships for various output settings. A typical output characteristic for a TAGS power unit is shown in fig. 4.6.

With no load across the output, the power-supply unit is not delivering current and the voltage is higher than it is during welding. This is the *open-circuit voltage* (o.c.v.) and is usually in the range 50 to 80 V. Once the arc is drawing current from the power supply, the voltage falls and over the operating range of 10 to 30 V the current varies to only a small extent.

Power-supply equipments with this type of output are known as 'drooping-characteristic' units, and a series of output curves is produced by changing the current setting (fig. 4.7). Sometimes such units are referred to as 'constant-

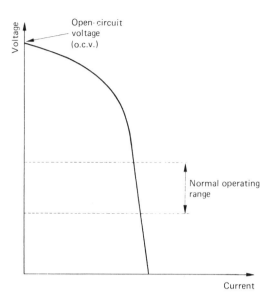

**Fig. 4.6**  Typical output characteristic for a power-supply unit used in TAGS welding

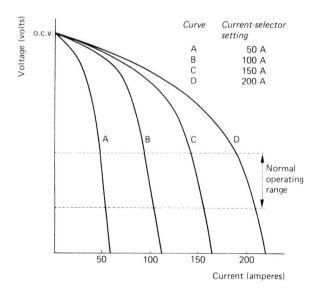

**Fig. 4.7**  Output characteristics for various settings of current

current' machines, but this is not strictly a correct description since the output shows some change of current with voltage.

If the arc and power-supply-output characteristics are plotted on the same graph, their intersection gives the working voltage and current. Taking the example given above for the welding of copper, the output for a 150 A setting on the power-supply unit is shown in fig. 4.8, together with the arc characteristics for arc lengths of 1, 2, and 3 mm. With a 2 mm arc the intersection is at 15 V for 150 A (point A), but if the welder increases the arc length to 3 mm the operating point now moves to the intersection of the 3 mm arc characteristic and the 150 A output line, i.e. 16.5 V, and the current level falls to about 143 A (point B). Conversely, if the welder shortens the arc length to say 1 mm, the operating point moves to point C and the voltage falls to 13.3 V with a corresponding current of 156 A. These results can be summarised:

| Arc length (mm) | Voltage (volts) | Current (amperes) | Power input (watts) | Variation in power (%) |
|---|---|---|---|---|
| 1 | 13.3 | 156 | 2074 | −7.8 |
| 2 | 15.0 | 150 | 2250 | 0 |
| 3 | 16.5 | 143 | 2359 | +4.8 |

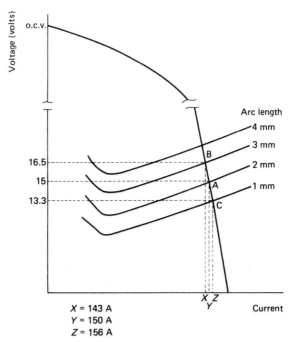

**Fig. 4.8** Variations in voltage and current with change in arc length (TAGS welding)

The important point to recognise is that, as the welder makes small changes in the arc length, possibly as a result of involuntary hand movements and so on, the power input remains within ±8% of the preset value, which is much better than the accuracy to which most welders can control travel speed.

## 4.5 Arc-length control in MMA welding

The above analysis can be applied to manual metal arc welding, but there is an additional requirement since the electrode is being melted. The arc length will therefore get longer unless the welder moves the holder towards the weld, i.e. feeds the electrode into the arc area, at a speed which matches the rate at which the electrode is shortening. This latter is known as the *burn-off rate*, and the volume of metal melted each second is dependent principally on the current – the higher the current the faster the burn-off rate, and vice-versa. In MMA welding, the consistency of the weld is therefore almost entirely dependent on the welder's skill in estimating the arc length and manually adjusting the electrode feed rate.

## 4.6 Self-adjusting arcs in MAGS welding

With MAGS welding the situation is in a sense the reverse of that described for MMA welding above, since the voltage setting of the power-supply unit, not the welder, governs the arc length. How then is the arc controlled?

Firstly we must recognise that with the smaller diameter wires used in the MAGS system the burn-off rates are far higher than in MMA welding and they vary much more with current. With only a small fluctuation in current, a significant change in burn-off rate is produced. Typical burn-off curves for low-carbon-steel wires in carbon-dioxide shielding are given in fig. 4.9.

An important point to note is the effect of electrode diameter. At 200 A the burn-off rates are 2.5, 5.1, and 10.4 m/min for 1.6, 1.2, and 0.8 mm diameters respectively. Equally significant is the change in burn-off rate produced by a given change in current – at 220 A the rates are 2.8, 5.6, and 11.5 m/min. Used in conjunction with a power-supply unit which keeps the voltage reasonably constant over a wide current range, this feature of MAGS welding provides the basis for the control of the arc.

A suitable output characteristic for a MAGS power supply is shown in fig. 4.10. There is a straight-line relationship between voltage and current, with the voltage falling by about 2 V for every 100 A increase in current. This type of supply is often referred to as 'flat' or 'constant potential'.

Consider an arc operating at 300 A, 35 V (point A on fig. 4.10), when the corresponding burn-off rate would be 7.6 m/min. Suppose the arc length increases, with the result that the voltage rises to, say, point B. This is accompanied by a decrease in current, giving a correspondingly lower burn-off rate (6.3 m/min). In the MAGS system the wire feed speed is constant, so we now have a situation in which the wire electrode is feeding into the arc area at a faster rate than it is burning back, since the burn-off rate has been reduced by the fall in current. The tip of the electrode approaches the surface of the weld pool and the arc length shortens, which means that the voltage returns to the original value. When this has been reached, the

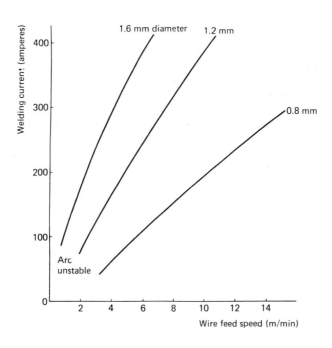

**Fig. 4.9.** Relation between wire feed rate and current (i.e. burn-off curves) for three diameters of steel electrodes in $CO_2$ welding

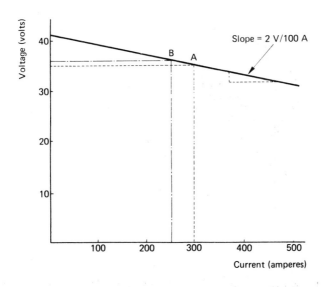

**Fig. 4.10** Output characteristics for a constant-potential power-supply unit

57

current will have risen and the burn-off rate once again matches the wire feed rate. The system has been returned to equilibrium. Conversely, if the arc length shortens, the voltage falls, the current rises, and the burn-off rate is increased. The wire, therefore, melts off faster than it is being fed into the arc area and the arc length is continuously increased until it returns to the preset value.

This sequence of events is known as self-adjustment, and the key to its effective operation lies in the magnitude of the change in burn-off rate. Suppose, once again, the arc length has been increased and the current has changed from $I_1$ to $I_2$ (fig. 4.11). Although the burn-off rate, corresponding to $I_1$, is now $B_2$, the wire is still feeding at the speed of the original

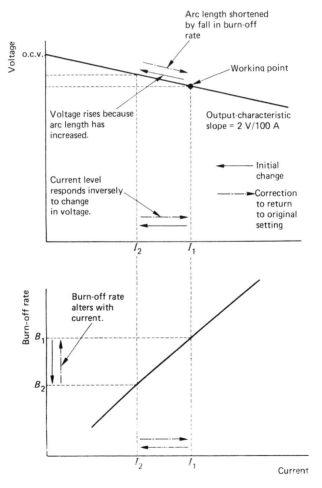

**Fig. 4.11** Changes in voltage, current, and burn-off rate with change in arc length (MAGS welding)

58

burn-off rate $(B_1)$, with the result that the tip of the wire is advancing towards the weld pool at a speed equal to $(B_1 - B_2)$. Hence, as the change in burn-off rate gets larger for a given difference in current $(I_1 - I_2)$, the speed with which the electrode tip returns towards its original burn-off position will increase.

With electrode wires in the range 0.8 to 1.6 mm diameter, this requirement for rapid self-adjustment is readily met. Thus, for example, with a 1.2 mm diameter steel wire in carbon-dioxide shielding the change in burn-off rate produced by a 20 A change in current is 0.51 m/min, which means that a 1 mm change in arc length is corrected in about $\frac{1}{510} \times 1$ min, i.e. 0.13 second. Again, with an 0.8 mm diameter wire the change in burn-off rate for 20 A change in current is 1.02 m/min, corresponding to a correction time of approximately 0.059 s per 1 mm change in arc length.

Pursuing this line of thought, we can readily see that with manual metal arc welding the correction times would be long because the changes in burn-off rates are small. For example, a 4 mm electrode operating at 200 A melts off at about 0.25m/min and a 20 A change in current produces only 0.02 m/min change in burn-off rate. An alteration of 1 mm in the arc length would therefore need about 3 seconds to be corrected by self-adjustment, which is much longer than it would take the operator visually to recognise and correct the change. It follows that, for manual metal arc welding, better results are obtained if the current is kept constant by the use of a drooping-characteristic power supply and the welder is left to control the arc length as described above.

### 4.7  Summary of heat-input control in arc welding
Summarising the control of heat input in arc welding, there are five main parameters which must be considered:

a) arc length,
b) arc voltage,
c) electrode feed rate,
d) arc current,
e) travel speed.

In the three manual arc-welding processes we have considered, the travel speed is under the direct control of the welder and the degree of consistency depends critically on individual skill. The way in which the other four parameters are controlled depends on the process and can be summarised as shown in Table 4.1 (page 60).

### 4.8  Power-supply units for arc welding
The power-supply unit in an arc-welding system must

a) isolate the secondary (or welding) circuit from the mains supply,
b) provide the voltages required for the welding operation,
c) supply current at the selected level,
d) have an output characteristic which matches the arc system,
e) incorporate a low-voltage supply for the operation of auxiliary units.

**Table 4.1**  Controller of welding parameters in TAGS, MMA, and MAGS welding

| Process | Arc length | Voltage | Electrode feed | Current |
|---------|-----------|---------|----------------|---------|
| TAGS welding | Welder | Welder | Not applicable | Power supply |
| MMA welding | Welder | Welder via arc length | Welder | Power supply |
| MAGS welding | Power supply via voltage | Power supply | Wire feed | Electrode speed via wire-feed motor |

Power units used in shops normally operate from a 440 V mains supply. With work on site, if connections to an electricity supply cannot be made, each unit can be self-contained with a petrol or diesel engine driving a generator or alternator.

*Alternating-current supplies*
Alternating current (a.c.) is widely used in MMA welding of steel and for the TAGS welding of aluminium and its alloys. It is rarely used in practice with MAGS techniques, in spite of the fact that it is technically feasible to operate consumable-electrode gas-shielded arcs with a.c.

When a.c. is required for welding, the power-supply unit is almost always a transformer, although suitable alternators are available. A typical equipment consists of two principal parts: the transformer itself and a means of adjusting the current.

The transformer has a single primary winding, tapped to accommodate different input voltages. It can be connected to a single-phase line or across two phases of a three-phase mains supply. This places an unbalanced load on the mains, and in large shops it is often necessary to make special provisions to distribute the load more uniformly across the three phases. The current output can be adjusted either by varying the inductance or by altering the magnetic coupling between the primary and secondary windings of the transformer.

An inductor placed in the line from the transformer to the electrode reduces the flow of the alternating current. The extent of this reduction depends on the amount of inductance in the circuit, and for current control during welding a means of altering this inductance must be provided. Three types of reactor are commonly used in power-supply units to give inductive control of current: tapped reactors, moving-core reactors, and saturable reactors.

**Tapped reactors**   These consist of copper cable wound on to a laminated core. Tappings are connected to the windings so that sections of the reactor can be taken out of circuit as required (fig. 4.12). Coarse and fine controls are provided, but only a limited number of settings can be accommodated.

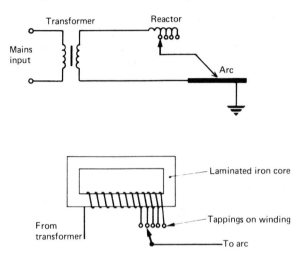

**Fig. 4.12**  Tapped reactor used to regulate welding current

**Moving-core reactors**   In these, a laminated core is moved into or out of a coil, thus increasing or reducing the inductance of the winding (fig. 4.13).

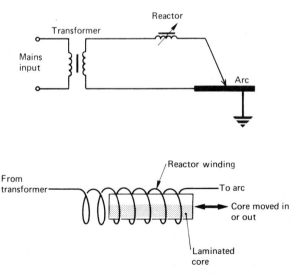

**Fig. 4.13**  Moving-core reactor used to regulate welding current

Although this involves the use of a slightly more complex mechanism, the system has the advantage of continuously variable adjustment.

**Saturable reactors**    These incorporate a second winding around the core (fig. 4.14). Direct current (d.c.) supplied to this winding affects the impedance offered to a.c. flowing in the main coil. Thus, by altering the d.c. in the control winding, the welding current can be regulated. In addition to providing continuously variable adjustment, saturable reactors can be remotely controlled. They are, however, appreciably more expensive than tapped reactors.

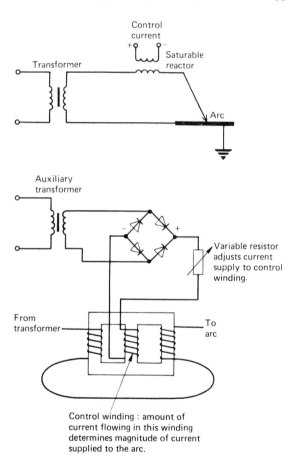

**Fig. 4.14**    Saturable reactor used to regulate welding current

As an alternative to the use of a reactor, the magnetic coupling between the two windings of the transformer can be altered to vary the current output. In commercial equipment this is done in three ways:

a) by using tappings on the primary or secondary windings;
b) by moving one coil along the core to change the spacing between the
   primary and secondary windings (fig. 4.15);

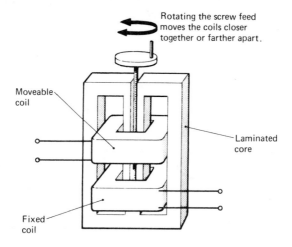

Rotating the screw feed
moves the coils closer
together or farther apart.

Moveable
coil

Laminated
core

Fixed
coil

**Fig. 4.15** Regulating welding current by altering the spacing between the
primary and secondary windings of the transformer

c) by keeping the windings in fixed positions with relation to each other, but
   moving the core in or out of the coils to increase or decrease the current
   (fig. 4.16).

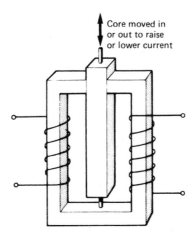

Core moved in
or out to raise
or lower current

**Fig. 4.16** Moveable-core transformer

All these designs provide good control of current and a suitable output for MMA and TAGS welding; the choice is determined mainly by cost and individual preference. An exception is the use of multi-operator sets where one transformer provides three or six outlets (fig. 4.17) – the current flowing in each secondary circuit must be controlled independently, and a separate reactor must be included in each lead.

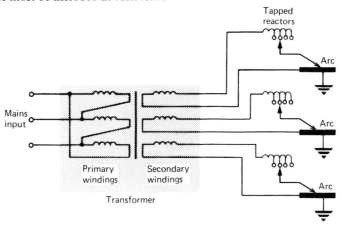

**Fig. 4.17** Multi-operator transformer unit

### Direct-current supplies – generators

For many years, motor-driven generators were the most commonly used supply units for welding with d.c. In recent years, they have been replaced in many fabrication shops by rectifier units, as the latter are quiet in operation and require less maintenance. Generators are still preferred for site work and for some shop applications, for example welding large-diameter pipes where independent control of voltage and slope is required. They can also be useful when mains-voltage fluctuations are troublesome.

The generator consists of an armature rotating in a magnetic field produced by coils which are connected in series and in parallel with the armature winding. The output is regulated by adjusting the current flowing in the series and shunt (i.e. parallel) windings. The armature must rotate at constant speed and is coupled to a motor unit. An electric motor is generally the first choice where a mains supply is available. In the absence of electrical power, the welding generator can be driven by a governed petrol or diesel engine.

### Direct-current supplies – rectifiers

In a rectifier unit, the a.c. output from a transformer is fed to a full-wave rectifier which converts it to d.c. (fig. 4.18). If a single-phase input is used, the d.c. has a pronounced 100 Hz ripple and for most applications some form of smoothing is required. A three-phase input is usually preferred, as this imposes a more uniform load on the mains supply – it has the added advan-

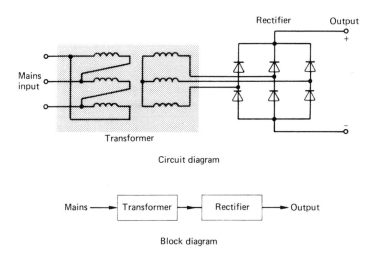

Circuit diagram

Mains ──▶ Transformer ──▶ Rectifier ──▶ Output

Block diagram

**Fig. 4.18** Simple three-phase full-wave rectifier unit for welding

tage of reducing the ripple on the d.c. output, thus removing the need for a smoothing circuit.

For MAGS welding, the transformer winding is tapped so that the output voltage can be selected to suit the arc length. Since there is no requirement for current control, the power-supply unit consists simply of the transformer and a rectifier.

In the case of MMA or TAGS welding, a drooping-output characteristic is required. This is obtained by inserting a reactor into the a.c. line between the transformer and the rectifier (fig. 4.19). The reactor behaves in a similar manner to that used in a.c. supply units. Most modern units use saturable reactors, since they are better suited to three-phase operation and give a remote-control facility. It should be noted that a reactor offers impedance only to a.c. – placed in the d.c. output from the rectifier, a reactor has little or no effect on the steady flow of current. It does, however, oppose any changes in current level, slowing down the rate of rise or fall. This feature is utilised in low-current MAGS welding (see chapter 6).

By providing extra connections to the output from the reactor in a transformer–rectifier set, it is possible to produce a combined a.c./d.c. unit suitable for MMA and TAGS welding. This type of supply is of considerable attraction where there is a mix of work, but it does cost more than the equivalent individual a.c. or d.c. unit. Combined a.c./d.c. sets usually have a single-phase input, as the a.c. output must be at 50 Hz.

*Operating problems with power-supply units*
The power-supply units discussed so far all have the merit of being relatively simple and reliable. To a large extent they have met the requirements of industry, but they do have deficiencies or drawbacks. Fortunately, these can

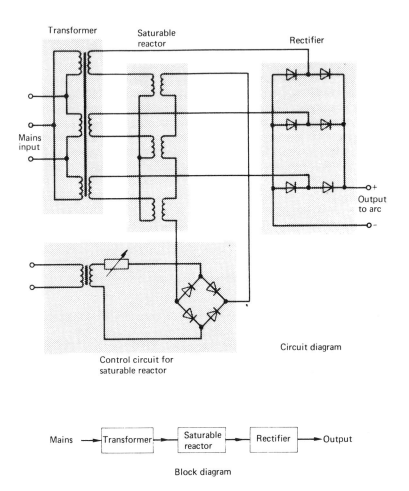

Fig. 4.19 Use of a saturable reactor to control current and provide a drooping-characteristic output from a rectifier unit

to a large extent be offset by a combination of the skill of the operator and the tolerance of minor changes in voltage and current which is a feature of arc-welding systems. The deficiencies become more apparent and start to pose problems as we look towards higher quality, greater consistency, and the use of automated plant. We will now identify the nature of the problems and see how power-supply units have been developed to meet present needs.

Firstly, we have to recognise one of the most common operating problems: mains fluctuations. Although we talk about the mains voltage as if it were a fixed value, under normal circumstances in a welding shop the voltage at the input to the welding power unit can be anywhere between 390 and 440 V. A number of factors can affect the mains voltage, all outside the control of the welder.

From the point of view of the welding operation, the significance of these variations is that they are immediately reflected in the arc behaviour. With TAGS and MMA welding the welder can readily adjust the arc to offset the effect of the change in arc voltage; but the current also alters, causing problems of, say, burn-through in a root run. In MAGS welding, on the other hand, a drop in the input voltage can result in a shift in the working point on the static characteristic (see page 58) which is not corrected by self-adjustment and continues until the mains voltage corrects itself. This could affect both the width of the weld and the penetration.

Secondly, the current-control systems which we have discussed do not have the ability to produce rapid changes. In later chapters we will look at techniques such as pulsing which require the current to be changed from a high to a low level 50 to 100 times per second. By their very design, the inductive devices used in conventional power units resist changes in current and it is difficult to achieve rapid response.

Thirdly, it is difficult to provide remote control on most of the power units we have studied in this chapter. The exception is the use of a saturable reactor, but, as we have seen, this is expensive.

Finally, the units tend to be large and heavy. They are not readily manoeuvrable. Most of the weight is in the transformer and inductors, if they are fitted. Other components such as rectifiers, relays, and fans are bulky rather than heavy.

The development of solid-state devices, such as the transistor, has provided an opportunity to radically redesign power-supply units so that the above problems can be overcome.

### Thyristor-controlled rectifiers

The rectifier in a conventional welding power unit is essentially a static component. It simply accepts the alternating current from the transformer or saturable reactor and converts it to d.c. without exercising any control. If a thyristor is used in place of the conventional rectifier, however, the current flow can be altered in response to a remotely controlled command signal (fig. 4.20).

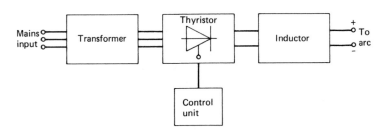

**Fig. 4.20** Thyristor-controlled welding rectifier

A thyristor is a solid-state device which allows current to flow in one direction, i.e. it behaves as a normal rectifier. However, in the appropriate half cycle it will conduct only after being 'fired' by a voltage pulse applied to the control terminal. Current will then flow for the remainder of that half cycle, until its direction reverses and the thyristor ceases to conduct. In a typical a.c. waveform, the mean current depends on the time during which current is flowing. Thus, if the firing of the thyristor is delayed after each time the voltage changes from positive to negative and vice versa, the current flows for a shorter time in each half cycle and the mean current is reduced (see the diagram for the primary circuit in fig. 11.7). The significance of this is that we can use the firing point of a thyristor to adjust the output from a welding power unit. In this way, thyristors can do the jobs of both the rectifier and the saturable reactor.

The main drawback of the thyristor is the very pronounced ripple in the output current. This can be reduced by including an inductor in the circuit.

Thyristor-controlled rectifier units are well suited to MMA and TAGS welding, where the facility of accurate remote control is of value.

### Transistorised units

The word 'transistor' is now well known thanks to the use in everyday life of transistorised radios, TV sets, and cassette players. Perhaps the most noticeable feature of these is their compactness. In welding power units, it is the way in which transistors operate that has opened up a totally new approach. Transistors can be used either as amplifiers or as switches.

When it is used as an amplifier, the transistor regulates the flow of current in response to a command signal. At first sight, it might be thought that this is similar to the thyristor unit, but the difference is very significant. In the case of the transistor, the current flows throughout the half cycle but its amplitude is reduced. This gives an output current which has less ripple than that from the thyristor unit.

A considerable amount of heat is generated in the transistor, and cooling is essential. There is also a limit to the amount of current that one transistor can carry and, for the current levels in welding, a large number of transistors are wired in parallel. Often there can be as many as 100 to 150 transistors in a pack. These power units are bulky and are very expensive, but they give very accurate control of the welding current and voltage.

The alternative is to use the transistor as a switch. In its simplest form, a bank of transistors can be used to switch resistors in or out of the circuit, thus altering the current (fig. 4.21). This is a useful technique at the low currents used for welding thin foil — say up to 10 A — but at high currents the resistors are expensive and need considerable cooling. A more successful technique is to remove the resistors and use the transistor as a rapid switch. The switching frequency is about 5 kHz, and the mean current output depends on the ratio of 'Off' to 'On' time. Heating in the transistors is less of a problem, as there is some cooling in each 'Off' period.

Perhaps the greatest attraction of transistorised units is the ability to monitor the output to the arc and make adjustments via the command signal

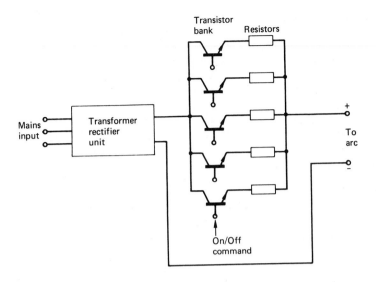

Fig. 4.21 Switched transistor and resistor control of current

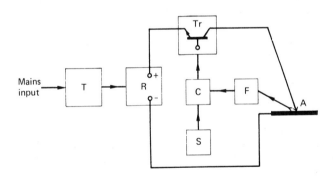

Fig. 4.22 Elements of a transistorised power-supply unit to give either a drooping characteristic or a constant-potential output

T—transformer   R—rectifier   Tr—transistor regulator   A—arc
F—feedback voltage and/or current from arc   S—reference setting
C—command unit (compares signals from F and S; amplifies error
to give command signal for Tr)

to the transistors (fig. 4.22). In this way the effects of mains-voltage variations can be offset. It is also possible to select either a drooping-characteristic or a flat-characteristic output from the one unit. Last and by no means least, the output can be programmed — a facility which is of great value in robotic applications (see page 163).

## Inverter systems

Although transistorised units solve the first three problems outlined above –
i.e. mains variations, slow response, and difficulty of remote control – they
are still of about the same size as or even larger than conventional units. They
are also appreciably more expensive and are most likely to be found in
specialised applications.

The answer to the weight problem lies largely in replacing the transformer.
This has been achieved in a new range of power units by using an inverter to
convert the mains frequency from 50 Hz to between 5 kHz and 25 kHz.
Transformers for currents operating at these high frequencies are much
smaller than those designed to handle the normal mains frequency of 50 Hz.

In a power unit for welding (fig. 4.23), the a.c. mains input is first rectified
to give d.c. This is fed to the inverter which converts it back to a.c. but at a
frequency of 5 kHz. It can now be reduced to the welding voltage by a small
lightweight transformer before it is rectified again to d.c. for the arc. As with
transistorised units, the output can be monitored and feedback used to
control the output.

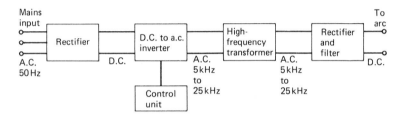

**Fig. 4.23** Welding rectifier with inverter

The use of inverters enables very compact units to be made. An inverter
set for welding up to 130 A is shown in fig. 4.24 – it weighs only 8 kg and
can be carried over the shoulder.

**Fig. 4.24** Portable inverter power unit

# 5 Operating characteristics of manual metal arc (MMA) welding

Undoubtedly the most widely used arc-welding process is MMA. The plant is relatively cheap and simple, the welder has considerable freedom of movement, since the point of welding can be as much as 30 metres from the power-supply unit, and it is possible to weld a variety of metals with the same equipment simply by changing the electrode type.

## 5.1 Welding current
Either direct or alternating current can be used for MMA welding. To some extent, the choice is based on experience and individual preference, but a number of factors need to be considered:

a) Virtually all MMA electrodes work on d.c., but only certain flux compositions give stable operation with a.c.
b) Transformers are easier to maintain than the generators or rectifiers used for d.c. Also, a.c. units are more robust.
c) D.C. arcs may be deflected from the joint by the magnetic effects produced when the welding current flows through the work (fig. 5.1). This phenomenon is known as arc-blow and is less common with modern electrodes than it used to be. It can, however, sometimes lead to difficulties. Arc-blow does not occur with a.c., as stable magnetic fields are not established.

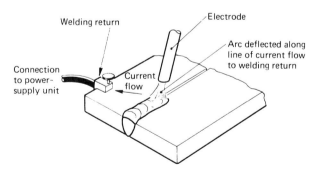

Welding return

Electrode

Connection to power-supply unit

Current flow

Arc deflected along line of current flow to welding return

**Fig. 5.1**    Arc-blow in MMA welding with direct current

d) Higher open-circuit voltages are required for a.c. The arc is extinguished each time the current goes through zero as the polarity is reversed (i.e. every one-hundredth of a second), fig. 5.2. If the weld pool is to remain

72

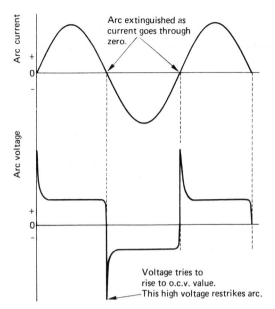

Fig. 5.2  Current and voltage waveforms in a.c. welding

molten, the arc must be instantaneously re-ignited. This requires that a voltage in excess of 80 V is applied to the electrode each time the current falls to zero. These high voltages can constitute a safety hazard (see appendix A) and d.c., with its lower o.c.v. of about 60 V, is often preferred for this reason alone. It is expected that modified power-supply units and new types of flux covering will be available in the future to enable a.c. to be used without the need for high voltage.

## 5.2 Position of welding

Ideally, the work should be positioned during welding so that the molten weld metal is held in place by gravity. This is called the flat position (fig. 5.3) and gives the welder the most favourable conditions for controlling the weld pool. It also enables high currents to be used, leading to faster welding. This implies that the work can be turned or manoeuvred easily. Many fabrications do not lend themselves to this treatment, and much of the welding in industry is done 'in position'.

Three main positions, in addition to flat, can be identified: horizontal, vertical, and overhead. There is a subdivision of the horizontal position, known as the horizontal–vertical, which relates specifically to 'T' joints in which the axis of one member is vertical while the other is horizontal. In all these positions the metal tries to run out of the joint under the effect of gravity, and the welder's technique must be modified to combat this tendency. A major contribution is made by the flux, and this will be discussed in more detail later. The welder controls the weld by lowering the heat input

73

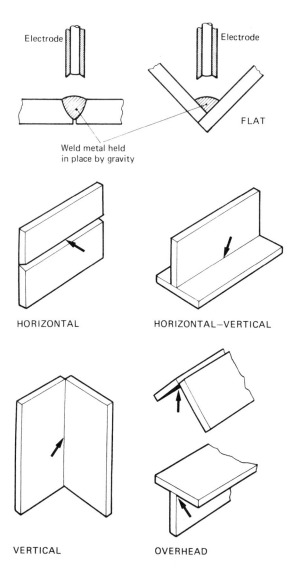

**Electrode**

**Electrode**

**FLAT**

Weld metal held
in place by gravity

**HORIZONTAL**

**HORIZONTAL—VERTICAL**

**VERTICAL**

**OVERHEAD**

**Fig. 5.3** Positions of welding

to reduce the fluidity and to give a small pool which solidifies before it has
time to run out of the joint. At the same time, the direction of the arc, i.e.
the angle between the electrode and the weld surface, can be varied to position
the weld pool to the best advantage.

The maximum current is lower in positional welding. Whereas 350 A can be
readily used for joints in the flat position, the welder would have considerable

difficulty in working above 160 A when welding overhead. It follows that the sequence used to deposit a weld of a given size differs from one position to another. This can be illustrated by considering the deposition of a typical fillet weld. The size of a fillet weld can be specified in a number of ways. This aspect is discussed in greater detail in chapter 7, but for our present purpose we will specify a leg length of 10 mm. This is the distance from the root of the weld to the toe and is, in effect, one side of the triangle formed by the fillet (fig. 5.4).

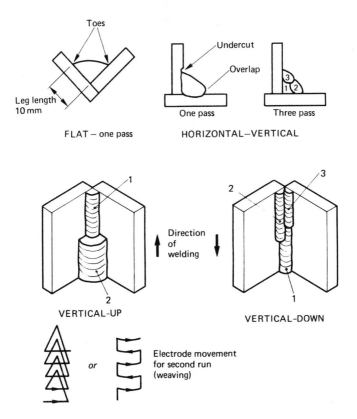

**Fig. 5.4**   Deposition of 10mm-leg-length fillet weld

In the flat position, with a current of 300 A the weld could be deposited in one pass. The welder moves the arc from side to side to ensure that the faces of the joint are properly fused. This is known as weaving and it produces a smooth flat surface. Using this technique, a one-metre length of weld could be deposited in about 10 minutes.

If the joint is in the horizontal–vertical position, attempts to deposit the weld in one pass result in a large uncontrollable pool which tries to flow on to the surface of the horizontal plate. The resultant weld is misshapen and there is overlap or lack of fusion at the toe. To obtain a uniform shape, the weld is deposited in three runs, using a lower current of about 200 A. Although the travel speed for each individual run is faster, the total time for the joint is 15 minutes, i.e. 50% longer than in the flat position.

When the joint is positioned so that the weld is deposited overhead, the current is reduced to 160 A and four or five runs may be needed to obtain the required size. The time has now increased to about 24 minutes.

When the plates are arranged so that the joint line is vertical, the weld can be started at the bottom and the arc moved upwards. This is the conventional method of fillet welding a joint in this position and is often referred to as the vertical-up technique. Two runs are required at about 145 A and the welder weaves the electrode along a triangular path to achieve fusion of the joint faces. The weaving also helps to distribute the heat and control the weld-pool fluidity. If the welder experiences difficulty in preventing the surface from sagging at the centre-line, the current is reduced to about 120 A and the weld is deposited in three passes using a smaller weave. The weld times range from 20 to 25 minutes, depending on the current used.

Alternatively, a vertical-down technique can be used. Welding starts at the top of the joint and the electrode is pointed upwards at the deposited weld metal. The rate of travel is critical, as the molten metal must not run down the joint ahead of the arc – if it does, fusion of the parent metal may not take place. The weld pool must be relatively small, and there is little scope for weaving. This means that our 10 mm-leg-length fillet weld may need five or even six weld runs. On the other hand, travel speeds with the vertical-down technique are comparatively high, and the overall time for the joint is about 17 min/m. The two main drawbacks with this method of welding vertical joints are that there are only a small number of suitable electrodes and it requires considerable skill to produce welds which are free of lack of side fusion.

## 5.3 Functions of flux covering

The main reason for using a flux covering in MMA is to protect the molten metal from atmospheric contamination. At the same time, the flux fulfils a number of functions, all of which contribute to the success of the welding operation.

### *Weld-metal protection*

The flux melts in the arc, along with the core wire. It covers the surface of the molten metal, excluding oxygen and nitrogen. When it has solidified, the flux forms a slag which continues to protect the weld bead until it has cooled to room temperature. While it is important to ensure that a good flux/slag covering is achieved, the flux composition must be chosen with a view to slag detachability, as the slag must be completely removed before the next run is laid. Ideally we would like a slag which lifts off the completed weld by itself. This is difficult to reconcile with the need for the slag to adhere to the weld metal during the cooling period, to avoid entry of air which would oxidise the surface. Slag detachability is also influenced by compounds included in the covering to achieve other objectives. The result is always a compromise, and the ease with which the slag can be removed varies from one electrode type to another.

Additional protection against atmospheric contamination can be provided by including compounds which decompose in the heat of the arc to form gases. These replace air in the arc atmosphere, thus reducing the risk of oxygen and nitrogen absorption. They may be carbonates giving off carbon dioxide, or cellulose which produces an atmosphere of hydrogen and carbon monoxide.

### Arc stabilisation

Although we often talk of the need to ensure that the arc is stable, it is very difficult to define what we mean by this. From an operational point of view, the easiest explanation is that we want the conditions to remain constant along the joint, unless the welder decides to make a change to accommodate some variation in fit-up etc. This implies that the top of the arc should always be at the centre of the core-wire cross-section, and the arc column should be in line with the axis of the electrode. In a groove, such as a single-V joint, the arc must not move from one side wall to the other seeking the shortest path from the electrode to the work: it must stay firmly fixed in the direction dictated by the welder. At the same time, the end of the electrode must melt uniformly and the metal must be transferred to the weld pool without upsetting the stability of the arc.

Another aspect of stability is the ease with which the arc can be established at the start of a weld run or re-ignited at the beginning of each half cycle when using a.c. In both cases, the gas in the arc gap must be ionised rapidly and at the lowest possible voltage. Ionisation is facilitated by inclusion of titanium oxide, potassium silicate, and calcium carbonate in the flux. Silicates and oxides which have been added principally for other reasons can also act as arc stabilisers.

### Control of surface profile

To understand the action of the flux in controlling the profile of a weld bead, we must first look at the role played by the surface tension of the weld pool. If the surface tension is high, the molten-metal surface becomes convex. On a flat plate, the liquid pulls away at the edges, and the angle of contact between it and the solid surface approaches $90^{\circ}$. This is another way of saying that the molten weld metal does not wet the solid parent metal. At the other end of the scale, if the surface tension is low, the contact angle is small and the molten metal has good wetting characteristics, so the surface of the pool is very flat.

Neither of these extremes is desirable in welding. A very high surface tension not only gives poor profile but also prevents the metal from flowing uniformly in the root of a 'V' groove. On the other hand, a very low surface tension makes it difficult to control the profile and to restrict the size of the pool. There is also a risk that the weld metal will flow over the joint faces before the arc has had time to melt them. Once again a compromise must be reached for satisfactory practical use.

The surface tension of the pool in arc welding is controlled by the oxygen level of the weld metal. This, in turn, is determined by the oxygen content in the flux. If the oxygen in the weld is low, this element will be transferred from the flux until a stable value has been established. The more oxygen there is in the flux, the higher will be the level in the weld. The effect this has on surface tension, and hence on surface profile, can be seen by examining fillet welds deposited in the horizontal–vertical position (Table 5.1). A low-oxygen weld has a high surface tension, and vice versa.

Table 5.1   Effect of flux oxygen content

| Oxygen content | High | Medium | Low |
|---|---|---|---|
| Surface tension | Low | Medium | High |
| H–V fillet-weld profile | | | |

In some cases, surface profile is not a prime consideration in formulating a flux, and a less-than-satisfactory profile may have to be accepted. This is the case with fluxes containing appreciable amounts of calcium fluoride and carbonate. These are used to give better mechanical properties which are associated with low oxygen contents. As we have seen, however, this means that the surface tension is high and we will inevitably have a convex profile if we use electrodes covered with these fluxes.

## Control of weld metal in position

The slag can be used as a mould which helps to keep the weld metal in place while welding in position. Three physical properties of the liquid slag must be kept in balance. Firstly, it must have sufficient fluidity to flow freely from the root of the weld – to give good visibility and to avoid slag being trapped when the weld solidifies. The fluidity must not be too high, however, otherwise the flux runs off the face of the weld. The problem is eased if the surface tension is reasonably high, since this helps the slag to stay in place. Finally, the slag should solidify rapidly to form a solid barrier which restrains the tendency for the weld metal to run out of the joint.

*Control of weld-metal composition*

One of the outstanding advantages of MMA welding is that we can make adjustments to the composition of the weld metal by adding alloying elements to the flux covering. We have already noted that the oxygen content of the weld pool depends to a large extent on the amount of this element present in the flux. Similarly, if we add manganese to the flux, in the form of ferro-manganese, it will transfer into the weld. The actual amount which ends up in the weld depends partly on the concentration of manganese but also on the composition of the flux. For any given combination of flux and weld metal, alloying elements are distributed between the two in a more-or-less fixed proportion. The transfer can, of course, work both ways. If the flux or slag layer is low in manganese, this element transfers from the weld until the correct proportion is established. Thus elements can be both added to and taken from the weld simply by altering the flux composition.

A good degree of control can be exercised over this transfer mechanism. The thickness of the covering on the electrode can be maintained to within close tolerances. Hence the ratio of flux to core wire is reproducible and the amounts of alloying elements which need to be added to produce a particular weld-metal composition can be calculated by the electrode manufacturer.

Broadly speaking, there are three main aspects of weld-metal composition control which should be considered: alloying, deoxidation, and contamination.

**Alloying**  If a core wire is used which has the same composition as that desired in the finished weld, it is necessary only to ensure that there is no loss into the flux – there is no need to add alloying elements. This is common practice with non-ferrous metals and some stainless steels.

With electrodes for use on low-carbon, carbon, carbon–manganese, and low-alloy steels, alloyed core wires tend to be expensive and it is preferable to carry out the alloying in the weld pool. In other words, the core wire is a low-carbon steel, and alloy elements – e.g. manganese, chromium, and molybdenum – are added to the flux. This gives the electrode manufacturer a considerable degree of flexibility and enables a large range of electrodes to be produced with the same core wire. This is of particular advantage with alloy steels where small quantities of special-composition electrodes may be needed.

**Deoxidation**  If a molten weld pool in steel contains large amounts of oxygen, bubbles of carbon monoxide are formed which are trapped in the solidifying metal to form porosity. The gas arises from a reaction between the dissolved oxygen and carbon:

$$FeO + C \longrightarrow Fe + CO$$

The carbon monoxide is not soluble in steel and is released as a bubble. Quite apart from the formation of porosity, the carbon loss is undesirable as it has an influence on the strength of the weld. The reaction must be supressed, and in MMA welding this is achieved by adding a deoxidant to the flux.

A deoxidant is an element for which oxygen has a strong affinity. When a deoxidant is added to a weld pool, the oxygen combines with it in preference to reacting with carbon. An oxide is formed which floats to the surface of the pool and mixes with the slag. In MMA welding of steel, the most commonly used deoxidant is silicon, which is added to the flux in the form of ferrosilicon. The deoxidising reaction in the weld pool is

$$2FeO + Si \longrightarrow 2Fe + SiO_2$$

Other deoxidants are used with different metals; for example, when welding copper it is usual to add phosphorous or zinc to remove the oxygen. These are not necessarily introduced via the flux and may be included in the core wire.

**Contamination**   So far we have discussed the use of additions to the flux to achieve desirable improvements in the weld. We must also recognise that the flux can introduce harmful elements.

In chapter 7 we will be discussing the role of hydrogen in the formation of cracks in a welded joint. Hydrogen comes from the covering of the electrode, which normally contains both absorbed and chemically combined moisture. We can dry the electrode to reduce the level of absorbed moisture, but the amount of combined water present depends on the chemical compounds in the flux. Various ways have been used to measure the quantity of hydrogen present. The most useful technique for comparing electrodes is to deposit a weld run under controlled conditions and collect the hydrogen which diffuses out over a specified period of time (fig. 5.5). The result is

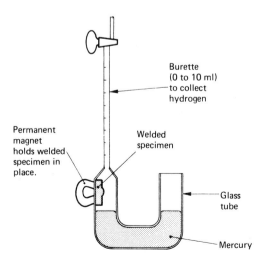

**Fig. 5.5**   Simple apparatus for collecting diffusible hydrogen

expressed as the millilitres of hydrogen in each 100 grams of weld metal; this is abbreviated to ml/100g. Data obtained from this test can be used in the design of weld procedures to avoid cracking (see chapter 7). The lowest hydrogen contents in MMA welding are obtained with electrodes containing large amounts of calcium compounds. These are discussed in more detail later, under the heading of 'Basic coverings' (page 82).

Other contaminants introduced via the flux covering are usually present by accident. Grease and oil can be collected during storage. Damp sulphurous fumes may be absorbed if electrodes are left in the vicinity of pickling vats. All of these are transferred with the flux via the arc to the weld pool, where they are absorbed. Careful storage and handling of electrodes are therefore of the utmost importance if high-quality welds are to be deposited.

## 5.4 Functions of individual constituents of a flux

It will be appreciated that a large number of chemical compounds are used in formulating a flux. Table 5.2 lists the compounds used in MMA electrodes for the welding of steels and indicates the roles which each flux constituent plays.

**Table 5.2** MMA electrode constituents and their functions

| Constituent | Primary function | Secondary function |
| --- | --- | --- |
| Iron oxide | Slag former | Arc stabiliser |
| Titanium oxide | Slag former | Arc stabiliser |
| Magnesium oxide | Fluxing agent | —— |
| Calcium fluoride | Slag former | Fluxing agent |
| Potassium silicate | Arc stabiliser | Binder |
| Other silicates | Slag formers and binders | Fluxing agents |
| Calcium carbonates | Gas former | Arc stabiliser |
| Other carbonates | Gas former | —— |
| Cellulose | Gas former | —— |
| Ferro-manganese | Alloying | Deoxidiser |
| Ferro-chrome | Alloying | —— |
| Ferro-silicon | Deoxidiser | —— |

*Note* Binders are used to give the flux covering mechanical strength and to help it adhere to the core wire. Fluxing agents are used to adjust surface tension and wetting characteristics.

## 5.5 Operating characteristics of electrodes for steels

There are four main groups of electrodes used in MMA welding of steels. They are distinguished by the major constituents of the flux covering which determine their operating characteristics.

**Acid coverings** are composed mainly of oxides and silicates and have a high oxygen content. They give smooth weld profiles with a tendency to concavity. The slag has a porous or honeycomb structure (inflated) and is easy to detach. Although they have good ductility, the welds tend to be low in strength, and for this reason acid electrodes are not widely used.

**Cellulosic coverings** have large quantities of organic material containing cellulose. Flour and wood pulp are common constituents. The organic compounds decompose in the arc to generate hydrogen which replaces air in the arc column. The presence of hydrogen increases the voltage across the arc and makes it more penetrating. For a given current, the depth of penetration with a cellulosic electrode is about 70% greater than with other types. Because most of the flux is decomposed, the resultant slag layer is thin and in some instances need not be removed before depositing the next weld run. The surface profile is poor and, while the mechanical properties are good, the hydrogen content is very high, which restricts the use of this type of electrode on high-strength steels.

**Rutile coverings** are based on titanium oxide. This compound has good slag-forming characteristics and produces a stable easy-to-use arc. Rutile electrodes are widely used and fulfil a general-purpose role in the fabricating industry. The deposits have a medium oxygen content (see page 78); hence surface profiles are acceptable, and slag detachability is good. By varying the additions of fluxing agents, the viscosity and surface tension can be adjusted to give electrodes which are suitable either for the flat position only or for use in all positions. Mechanical properties are adequate for most structural steels, but it is not easy to achieve high tensile strengths. The flux can be dried, but it will always contain a reasonable proportion of combined water which is needed to keep the covering intact. If this moisture is driven off, the binding of the flux will suffer. The hydrogen contents of the weld metal are high, being of the order of 25 to 30 ml/100 g. This is above the level normally considered desirable for high-strength steels.

**Basic coverings** mainly contain calcium compounds such as calcium fluoride and calcium carbonate. They are sometimes referred to as 'lime-coated' electrodes and are used principally for the welding of high-strength steels. The flux can be dried by baking at approximately $480^\circ$C, and, if the electrodes are stored at $150^\circ$C until used, the hydrogen content of the weld metal can be reduced to between 10 and 15 ml/100 g. At this level the risk of cracking in high-strength steels is minimised. Electrodes which can be treated in this way are called 'hydrogen-controlled'. The mechanical properties are very good, and it is possible to produce weld deposits which match almost all commercially used carbon–manganese and low-alloy steels. The oxygen content of the slag is low, however, and surface profiles of weld deposits are convex (see page 78). The slag is difficult to detach, and more fumes are given off than with other types of electrode.

**Iron-powder additions** are sometimes made to the flux covering to increase the electrode efficiency. This is defined as the mass of metal deposited as a percentage of the mass of core wire melted:

$$\text{electrode efficiency } \% = \frac{\text{mass of metal deposited}}{\text{mass of core wire melted}} \times 100\%$$

In general the efficiency of MMA electrodes is between 75 and 90%. Some of the metal is lost as fume and oxides in the arc, but the largest source of loss is spatter, i.e. droplets of molten metal ejected from the arc.

Electrode efficiencies rise above 100% if iron powder is added to the flux covering. Thus, in a deposit made with an electrode said to have an efficiency of 160%, slightly more than 60% of the weld metal will have come from the flux, the remainder being melted core wire.

In addition to giving high deposition rates, iron-powder additions to the coating also tend to increase the oxygen content of the slag and produce welds with smooth flat surfaces. The slag detaches readily, and the electrode is easy to use in the flat or horizontal–vertical positions.

### 5.6 Current ranges for MMA electrodes
The core wires of electrodes for MMA welding are generally 450 mm long and have a diameter within the range 2.5 to 6.3 mm. As the electrical connection is made at the top of the electrode, the full welding current flows along the core wire (fig. 5.6). This core wire has electrical resistance, and heat

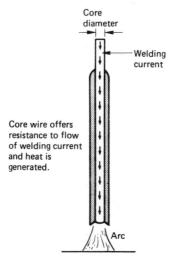

Fig. 5.6    Current flow in an MMA electrode

is generated by the passage of the current. If the temperature rises too high, there is a danger that the electrode will melt prematurely. Before this happens, however, the flux can be damaged. Moisture evaporated from the compounds in the flux causes it to flake off, leaving lengths of the core wire uncovered. At the same time, oxidation of some of the alloying elements can affect the composition of the resultant weld. With these points in mind, the electrode manufacturer stipulates a maximum current for each electrode diameter. There is also a lower limit of current, below which the arc becomes unstable. Table 5.3 gives typical ranges for MMA electrodes: the actual values depend on the type and composition of the flux and are specified by manufacturers for their own products.

Table 5.3    Current ranges for MMA electrodes

| Core-wire diameter (mm) | Current (amperes) | |
|---|---|---|
| | Minimum | Maximum |
| 2.5 | 50 | 90 |
| 3.2 | 65 | 130 |
| 4.0 | 110 | 185 |
| 5.0 | 150 | 250 |
| 6.0 | 200 | 315 |
| 6.3 | 220 | 350 |

## 5.7  Choosing and specifying an electrode

There is no single answer to the question 'Which is the best electrode?' Each application must be evaluated by examining a number of aspects; for example

a) what is the composition of the metal to be melted?
b) is there a risk of weld metal cracking?
c) what mechanical properties are required?
d) is a.c. or d.c. plant available?
e) what is the position of welding?
f) what is the thickness of the parent metal?
g) what is the type of joint?
h) are there any limitations on heat input?

Having answered these and other questions, an electrode type is selected which gives the optimum performance at an economic price. It will then be found that the type of electrode chosen is marketed by a number of electrode manufacturers. Each uses a brand name for the electrode, making it difficult to compare one product with another. The task of comparison is made easier, however, by use of an electrode-specification system. Most industrialised countries have produced their own system – some are based on the international standard ISO 2560; others are unique to a particular country. Mostly the index number allocated to an electrode indicates the properties of the weld deposit, the type of covering, and the operating characteristics (i.e. type of current, a.c. or d.c., and best positions of welding). This can be illustrated by two examples:

a) *United Kingdom: BS 639 (based on ISO 2560)*

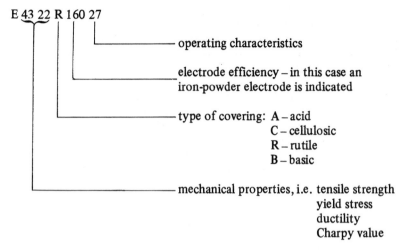

E 43 22 R 1 60 27

operating characteristics

electrode efficiency – in this case an iron-powder electrode is indicated

type of covering: A – acid
C – cellulosic
R – rutile
B – basic

mechanical properties, i.e. tensile strength
yield stress
ductility
Charpy value

*Note*  The letter H added to the end of the code number indicates a hydrogen-controlled electrode.

b) *United States of America: AWS–ASTM code*

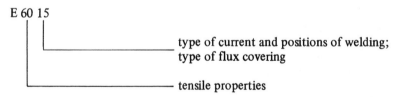

E 60 15

type of current and positions of welding; type of flux covering

tensile properties

The British Standard electrode-specification system is discussed further in appendix C.

# 6 Operating characteristics of gas-shielded welding

## 6.1 Tungsten arc gas-shielded (TAGS) welding

Perhaps the outstanding feature of TAGS welding is that the tungsten electrode is non-consumable. As there is no metal transferred across it, the arc is used solely to melt the parent metal. If additional metal is required to build up the weld surface, fill a groove, or form a fillet in a 'T' joint, it can be added to the weld pool separately in the form of filler wire. The degree of control which this arrangement gives the welder makes the TAGS process particularly suitable for the welding of sheet material. Welds can be deposited in all positions, but the level of skill required is very high. Control of the weld pool is also important in the deposition of root runs in pipe. TAGS welding is often used for this purpose in the fabrication of pipework for high-pressure steam lines and for chemical plant (fig. 6.1).

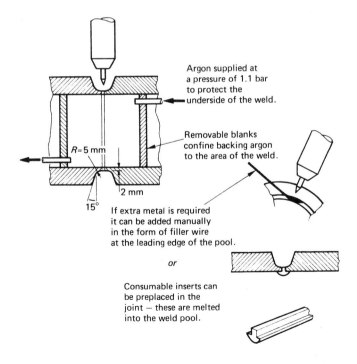

Argon supplied at a pressure of 1.1 bar to protect the underside of the weld.

Removable blanks confine backing argon to the area of the weld.

$R = 5$ mm

2 mm

15°

If extra metal is required it can be added manually in the form of filler wire at the leading edge of the pool.

*or*

Consumable inserts can be preplaced in the joint — these are melted into the weld pool.

Fig. 6.1    TAGS welding of root runs in pipe

*Welding current*

Direct current with the electrode connected to the negative pole of the power-supply unit is used for all metals except aluminium. This arrangement gives a stable arc. About 60% of the total heat is developed at the surface of the work, thus the larger proportion of the heat is available to melt the parent metal. Some 10% is lost by radiation and only 30% is released at the electrode; hence it is relatively easy to keep the electrode below its melting point (3370°C). The end of the electrode is tapered to encourage the arc to operate from the tip. This improves the arc stability and helps the welder to keep the arc length constant.

Aluminium and its alloys present a major problem. When the parent metal is melted, the oxide skin remains intact as its melting point is appreciably higher than that of aluminium. As there is no flux in TAGS welding, the oxide skin must be dispersed by the action of the arc. This happens only when the electron flow is from the weld pool, i.e. when it is negative and the electrode is positive. Under these conditions, oxide-free welds can be produced in aluminium. The main drawback is overheating of the electrode, which occurs because the biggest proportion of the heat is being developed in it, i.e. it is the reverse of the situation we discussed above with the electrode negative. This poses so many difficulties that electrode-positive d.c. is rarely used in practical applications.

The solution lies in the use of an a.c. arc. During the half cycles when the electrode is positive, the oxide is dispersed from the molten aluminium. The electrode cools during the negative half cycles while heat is built up in the weld pool. In this way a uniform distribution of heat is achieved and the electrode can be prevented from melting to any great extent – usually the tip becomes a hemisphere of molten tungsten (fig. 6.2).

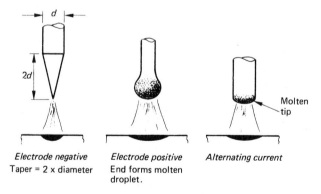

Electrode negative
Taper = 2 x diameter

Electrode positive
End forms molten droplet.

Alternating current

**Fig. 6.2**  Electrode configuration in TAGS welding

*Electrodes*

Although a pure tungsten electrode can give satisfactory results, it is usual to use an electrode which contains additions of either thoria (thorium oxide) or zirconia (zirconium oxide) to improve arc striking and stability. Thoriated electrodes are used for d.c. welding, while zirconiated electrodes are preferred for a.c. The diameter of the electrode is chosen to suit the current (Table 6.1).

**Table 6.1**  Typical maximum currents for TAGS electrodes

| Electrode diameter (mm) | Maximum current (amperes) | |
| --- | --- | --- |
| | Thoriated (d.c.) | Zirconiated (a.c.) |
| 1.2 | 70 | 40 |
| 1.6 | 145 | 55 |
| 2.4 | 240 | 90 |
| 3.2 | 380 | 150 |
| 4.0 | 440 | 210 |

### Starting the arc

By contrast with MMA welding, the arc cannot be started by touching the electrode on to the plate. In the TAGS system this would contaminate the tungsten with the parent metal and lower its melting point. A method of arc initiation must be incorporated which allows the welding current to flow across the gap between the electrode and the work. This requires ionisation of the gas in the gap, which can be achieved by injecting a high-frequency (h.f.) discharge or a high-voltage spark. An h.f. unit is essentially a tuned-oscillator circuit consisting of a transformer, a capacitor, and a spark gap (fig. 6.3). The tuned circuit operates at a frequency between 300 kHz and 3 MHz. The spark starter uses a high-voltage coil of the type found in car ignition circuits, with the arc gap replacing the sparking plug.

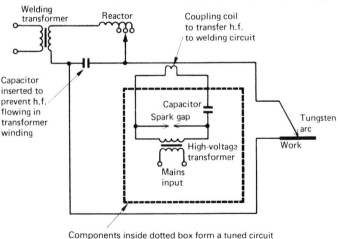

Components inside dotted box form a tuned circuit which generates a high-frequency wave.

**Fig. 6.3**  High-frequency arc-starter unit for TAGS welding

Once the gas in the gap has been ionised by either the h.f. or the spark discharge, the full welding current flows and the arc is established. With d.c. welding, the arc starter can be switched off at this point. On the other hand, in a.c. welding the arc is extinguished each time the current passes through zero going from negative to positive and vice versa. The arc must be re-ignited one hundred times each second. This means that the arc starter must operate continuously throughout the welding operation.

### *Rectification in a.c. arcs*
Ideally the current flow in the welding circuit must be the same during the negative and positive half cycles. If there is a difference, one of two problems can arise. A reduction in the current during the negative half cycle leads to overheating of the electrode. On the other hand, if the positive half cycle is restricted, the oxide on the pool is not adequately dispersed. In both cases, the result is similar to that which would be produced if a direct current was superimposed on the alternating current. For this reason the imbalance is referred to as rectification (fig. 6.4).

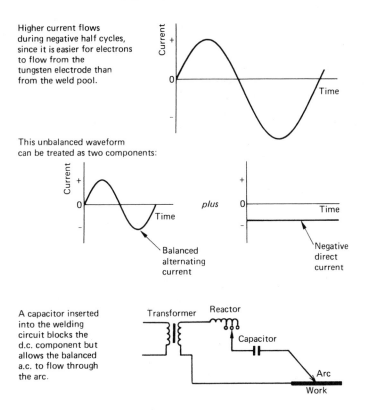

**Fig. 6.4**   Inherent rectification in TAGS welding with alternating current

89

In practice, reductions during the negative half cycles rarely occur. The most common form of rectification is that resulting from the difference between the arc voltages during the negative and positive half cycles. A higher voltage is needed when the electrode is positive, with the result that a lower current flows. This is known as 'inherent' rectification and is offset by including in the welding circuit a large capacitor which allows the a.c. to pass but blocks the d.c. component (fig. 6.4).

*Shielding gas*

In TAGS welding, pure argon is normally used as a shielding gas. It works well with both d.c. and a.c. and is compatible with all the weldable metals. The nozzles used are designed to match the flow characteristics of argon so that there is no turbulence (fig. 6.5). This gives protection to the top of the weld pool, but the underside in a root run remains exposed to air and can oxidise rapidly. Where this is a problem, a secondary supply of argon is provided on the reverse side of the joint.

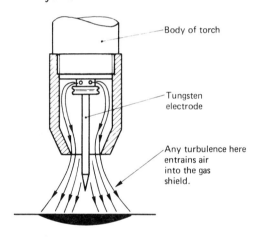

**Fig. 6.5** Gas shielding in TAGS welding

Helium is an alternative to argon which offers advantages when welding metals having a high thermal conductivity. The arc voltage is higher with helium, giving a greater heat input. The cost of helium prevents it being widely used, but it can usefully be mixed with argon when welding copper.

*Low-current operation*

TAGS welding is mainly used on sheet material less than 4 mm thick; above this, travel speeds become very slow. The lower limit of thickness is deter-mined by arc stability. At currents less than 10 A the arc wanders from point to point on the surface of the weld pool and drifts from the tip of the electrode. Stable operation is possible if the current is supplied in pulses. The pulse frequency varies from 10 per second to 1 per second (fig. 6.6). The current during the pulses is at a level where the arc is stable. The heat input depends on the mean current, which is controlled by the pulse height, pulse frequency, and pulse duration. The weld consists of a series of overlapping circular spots, each of which is melted by one pulse and freezes before the next pulse.

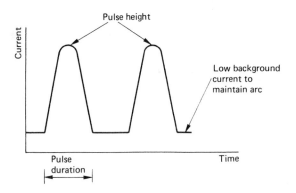

**Fig. 6.6** Pulsed current supply for TAGS welding

Pulsed TAGS welding can be used not only on thin sheet but also in root runs, as it gives better control of penetration than conventional continuous-current techniques.

## 6.2 Metal arc gas-shielded (MAGS) welding
In many ways MAGS welding is a direct competitor with the MMA process. It offers faster welding in similar applications, but uses more expensive equipment and consumables. The quality of the welds is comparable, and the choice between the two is often based solely on the relative operating costs. MAGS welding has opened up new areas of work in the sheet-metal industry for which MMA was not suitable. It plays an important role, for instance, in the production of car bodies, where freedom from frequent electrode changing and from the need to remove flux are important prodution factors. In thick-plate applications, the high rate at which electrode metal is deposited can help to reduce production costs.

### Electrode wires
Although the absence of flux is an advantage in the welding operation, it does mean that all the alloying elements and deoxidisers must be present in the electrode wires. Electrodes range in diameter from 0.8 to 1.6 mm and are specially produced for MAGS welding, as the surfaces must be free of contamination.

Steel wires contain manganese and other elements such as chromium, molybdenum, and nickel to give strength or corrosion-resistance to the weld metal. The amounts added are usually slightly larger than the corresponding parent-metal contents to allow for alloy loss in the arc. Silicon is used to deoxidise the weld pool as in MMA welding, but must now be included in the wire itself. Aluminium is added to some electrodes to act as a deoxidant. It reacts with oxygen at a faster rate than silicon but it can leave a detrimental effect on the ductility of the weld metal where the sulphur content of the parent metal is high enough to form aluminium sulphide in the weld. Silicon contents range from 0.4 to 0.9% and aluminium, if added, is usually about 0.15%. The wires are copper-coated to provide better electrical pick-up in the contact tube and to reduce the risk of corrosion of the wire in storage.

Electrode wires for stainless steel, weldable aluminium alloys, and other non-ferrous metals usually match the plate in composition. They are not copper-coated.

### Shielding gases

In contrast with TAGS welding, the shielding gas in the MAGS system is varied to suit the parent metal (Table 6.2). To some extent this is done to reduce costs, but more often it is to achieve better arc operation. Once again, the nozzles must be designed to give a non-turbulent flow of shielding gas over the weld area. The flow rates are significantly higher than with TAGS – 12 to 20 litres/min compared with 4 to 6 litres/min.

**Table 6.2**  Shielding gases for MAGS welding

| Metal | Shielding gas |
|---|---|
| Aluminium and alloys | Pure argon |
| Nickel and alloys | Pure argon |
| Copper | Argon + helium |
| Stainless steel | Argon + 5% oxygen |
| Low-carbon steel and carbon—manganese steel | Carbon dioxide or argon + 20% $CO_2$ or argon + 5% $O_2$ |
| Steels with 1 to 2% chromium | Argon + 20% $CO_2$ or argon + 5% $O_2$ |
| Steels with more than 2% chromium | Argon + 5% $O_2$ |

Pure argon is used for aluminium and its alloys, nickel and its alloys, and many non-ferrous metals. Helium added to the argon increases the heat input, and the mixture is used to advantage when welding metals with a high thermal conductivity, such as copper.

With ferrous metals, including stainless steels, a MAGS arc operating in pure argon tends to be unstable. The low oxygen content of the arc atmosphere also causes the surface of the pool to become convex and to pull away from the edges, leaving the weld depressed or undercut (see fig. 7.10). An addition of 1 to 5% oxygen to the shielding gas overcomes this problem. Subsequently the oxygen combines with silicon in the electrode wire to give a glass-like slag which floats to the surface of the molten pool.

Carbon dioxide is also used to shield the arc when welding low-carbon steel and carbon–manganese steels ($CO_2$ welding). Although this gas is chemically active, in the arc it breaks down to a mixture of oxygen and carbon monoxide. Extra silicon added to the electrode removes the oxygen, leaving the weld pool blanketed by carbon monoxide which does not react with molten steel. A mixture of argon and carbon dioxide in the ratio 4:1 is also widely used. The choice between argon + 5% oxygen, argon + 20% carbon dioxide, and pure carbon dioxide as a shielding gas for the welding of steels is far from simple and is based on a comparison of arc behaviour, spatter levels (i.e. the amount of metal ejected from the arc as molten droplets), weld profile and penetration, and cost. Even the cost aspect is not clear-cut, as it is affected by the type of supply, i.e. individual cylinders of compressed gas or large containers of liquefied gas (fig. 6.7), the quantity purchased, and whether the gas is supplied premixed or is mixed at the user's factory.

The use of carbon dioxide, either by itself or mixed with argon, is generally restricted to the welding of steels which do not contain chromium. Carbon dioxide reacts with chromium to form a carbide, thus reducing the effective content of this alloying element in the steel.


*Welding current*

For commercial applications, MAGS welding uses d.c. with the electrode positive. Two current ranges are normally used. The higher range covers from 300 A upwards and is confined to joints in plates in the flat and horizontal–vertical positions. The heat input is too high for positional work, and the penetration is too great for manual welding of sheet. These applications are better suited to the low-current range (50 to 180 A), where it is possible to produce a small weld pool which freezes quickly and has less penetration. Aluminium provides an exception to this in so far as positional welding is concerned. The high-current range extends from 220 A and can be used for welding vertically and overhead. The pool is retained in position by an oxide skin which forms over the surface. At the same time, heat is conducted away rapidly from the fusion zone, since aluminium has a high thermal conductivity.

The arc is very stable at high currents. Droplets are detached regularly from the end of the electrode and are transferred to the pool by electromagnetic forces acting in the arc along the line of the axis of the electrode

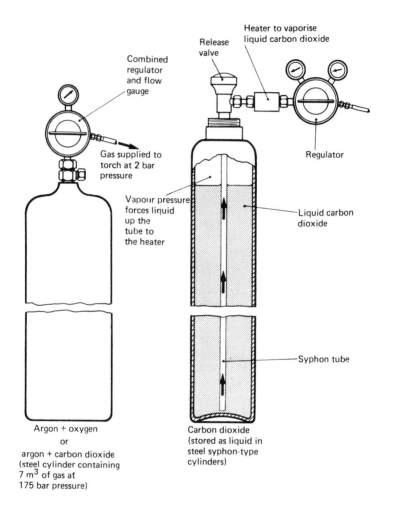

Combined regulator and flow gauge

Release valve

Heater to vaporise liquid carbon dioxide

Gas supplied to torch at 2 bar pressure

Regulator

Vapour pressure forces liquid up the tube to the heater

Liquid carbon dioxide

Syphon tube

Argon + oxygen
or
argon + carbon dioxide
(steel cylinder containing
7 m³ of gas at
175 bar pressure)

Carbon dioxide
(stored as liquid in
steel syphon-type
cylinders)

**Fig. 6.7** Cylinders used for argon and carbon-dioxide gases in MAGS welding

(fig. 6.8). This is known as *spray transfer*. In argon shielding, the droplets are detached completely from the electrode before they reach the weld surface. On the other hand, in carbon-dioxide shielding the droplet becomes very large before being detached and may actually touch the weld pool, creating a short circuit. It is the rapid rise in current during this short circuit which leads to smaller droplets being ejected from the arc area as spatter.

If we wish to weld in position or to join sheet metal, we must reduce the heat input. However, at low currents, i.e. less than 200 A, the forces in the

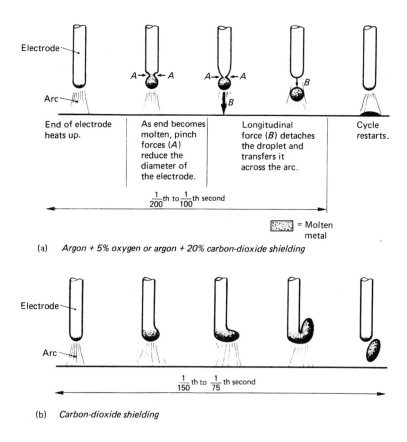

(a)   *Argon + 5% oxygen or argon + 20% carbon-dioxide shielding*

(b)   *Carbon-dioxide shielding*

**Fig. 6.8**   Spray transfer in MAGS welding

arc are too small to detach and transfer the droplets. The arc is unstable and, unless the mechanism of transfer is altered, the system cannot be used for welding. The problem can be solved in two ways. In the first, known as *dip transfer*, a short arc is established by lowering the voltage to about 21 V. At currents below 200 A, the end of the electrode melts slowly. As the electrode is being fed into the arc at a steady rate, the arc gap is shortened until the tip touches the weld pool (fig. 6.9). The output from the power supply is now being short-circuited, and a very high current flows through the electrode. If this were allowed to continue, the wire would melt and be ejected as spatter. The inclusion of an inductance in the welding circuit controls the rate at which the current rises, so that the end of the electrode is melted uniformly and flows into the weld pool, re-establishing an arc gap. This cycle is repeated 50 to 200 times each second.

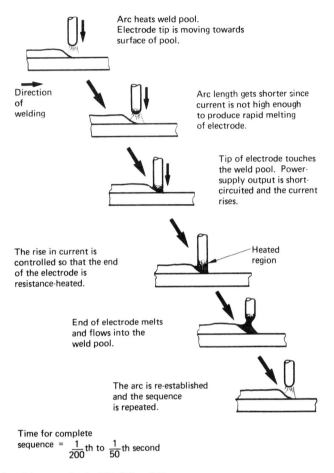

Arc heats weld pool.
Electrode tip is moving towards
surface of pool.

Direction
of
welding

Arc length gets shorter since
current is not high enough
to produce rapid melting
of electrode.

Tip of electrode touches
the weld pool. Power-
supply output is short-
circuited and the current
rises.

The rise in current is
controlled so that the end
of the electrode is
resistance-heated.

Heated
region

End of electrode melts
and flows into the
weld pool.

The arc is re-established
and the sequence
is repeated.

Time for complete
sequence = $\frac{1}{200}$th to $\frac{1}{50}$th second

**Fig. 6.9**   Dip transfer in MAGS welding

The alternative technique is known as *pulsed transfer* (fig. 6.10). The arc is operated with a normal arc length and a low background current of the order of 50 to 100 A. At this current, heat is maintained in the weld pool and some electrode heating takes place. Pulses of high current (in excess of 300 A) are superimposed on to the low background current. During each pulse, the arc behaves as if it were operating in the spray range and a single droplet is transferred before the current is returned to a low level. This gives spray-type transfer at a low heat input, because the latter is proportional to the mean current. The most commonly used pulse frequency is 50 Hz.

Both dip and pulsed transfer enable MAGS welding to be operated at low current. This means that positional welding and the joining of sheet material down to 0.8 mm thickness are possible. Dip transfer is by far the most commonly used technique but it is applicable only to metals which

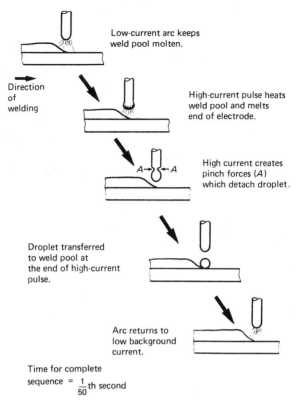

Low-current arc keeps weld pool molten.

Direction of welding

High-current pulse heats weld pool and melts end of electrode.

High current creates pinch forces (A) which detach droplet.

Droplet transferred to weld pool at the end of high-current pulse.

Arc returns to low background current.

Time for complete sequence $= \frac{1}{50}$ th second

**Fig. 6.10** Pulsed transfer in MAGS welding

have a relatively high electrical resistance, i.e. steels and some nickel alloys, since the electrode is melted by resistance heating during the short-circuiting period.

Improvements in the control and stability of both pulsed and dip-transfer systems are possible with solid-state power-supply units (see page 67). The output from these units can be controlled in response to feedback from the arc. This means that relationships can be established between voltage, current, waveform, and wire-feed rate, and automatic correction can be made during the welding operation. The name given to this is 'synergic-MIG'. Apart from ensuring greater consistency, 'synergic' techniques offer the possibility of producing 'one-knob' units which are easier for the welder to set up and use.

# 7 Weld quality and strength

## 7.1 Criteria for weld acceptability

Welds are not of themselves either good or bad, although sometimes examples may be produced which are so full of holes and inclusions that it is difficult to imagine an application for which they could be used. When we refer to a weld as being 'bad', we really mean that it is unacceptable for the purpose we have in mind and we must recognise that a standard or quality which one person might regard as reasonable may be too high or too low for other fields of work. Take, for example, a butt joint in a nuclear reactor. Once the reactor is working it might be possible, although undesirable, to carry out repairs to a weld which develops a fault. Of far greater importance is the fact that the failure of a critical joint could have disastrous consequences and it would be quite unjustifiable to risk putting into service a weld which contains any significant defects. In a storage tank, on the other hand, one of the main considerations would be the corrosion-resistance of the weld, and defects which would be a cause for immediate rejection in the case of the nuclear reactor could well be considered harmless in the tank, since high strength is not always a prime requirement for service.

While it is relatively easy to make general comparisons of the type given above, it is much more difficult to write a precise definition for the acceptable quality of a weld. Usually we attempt to relate this definition to the presence of defects such as porosity, undercut, lack of bonding between weld metal and plate, cracks, and pieces of slag trapped in the weld. It is unfortunate the term 'defect' has been used in this context, since it implies that the very presence of one or more of these makes the weld defective, but, as we have seen, there are cases where they can be tolerated. In fact the basis of a specification for weld quality is frequently an assessment of the number and size of the defects which can be present in a weld before it is considered to be defective and therefore rejectable. This is not to say that we should aim to produce welds containing defects – on the contrary, the welder should always use procedures and techniques which avoid the formation of the faults noted – but, because reality is not perfect, we know that defects will occur, and our standard of acceptance is simply stating the point at which corrective treatment must be applied.

The designer of a component or structure may have to take into consideration a number of criteria, such as:

mechanical properties,
resistance to corrosion,
resistance to oxidation,

resistance to abrasion,
leak tightness,
vacuum tightness,
stiffness.

The acceptance standard for the welds must reflect the significance of each defect in relation to those design criteria which are of importance. Mostly, however, we are concerned with how welds behave under various types of tensile loading.

## 7.2 Strength of butt-welded joints

The designer of a welded structure bases his calculations on the properties of the parent metal he has chosen. The assumption is made, therefore, that the weld will have properties at least equal to those of the parent metal. Where this is not so, there must be a clear understanding of the effect of welding on the strength of the joint, so that the designer can allow for this in the calculations. It may also be that with certain types of loading, for example fatigue, the presence of a welded joint automatically reduces the permissible maximum stress.

We can start with a butt joint in which there is bonding between the weld metal and parent plate at all points throughout the thickness. It is useful to make the proviso that the excess metal and penetration bead should be ignored, since we cannot control their dimensions precisely (fig. 7.1). The load-bearing thickness of the butt weld – the throat – is now equal to the thickness of the parent metal. We have two cases to consider, i.e. when the weld-metal yield stress is firstly greater and secondly less than the yield stress for the parent metal.

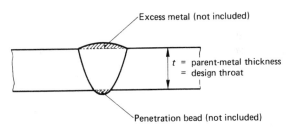

Fig. 7.1   Design throat in a butt joint

We can see the importance of the weld-metal yield stress by looking at the behaviour of a test-piece during a tensile test in which the test-piece contains a weld. The weld is transverse to the applied stress and has been machined flush with the surface of the test-piece. As the weld metal and parent material are essentially similar, they have the same modulus of elasticity and stretch uniformly as the load is applied in the elastic range. If the weld metal has the higher yield stress, yielding occurs first in the parent metal at some point away from the weld, accompanied by necking. The true stress in this region increases more rapidly than the stress in the weld metal, since the cross-section is being reduced by the necking, and fracture takes place in the plate. It will be seen that this is the basis for the often expressed view that a weld repair is stronger than the original part, but it will be realised that this applies only to tensile loading in those circumstances in which a weld metal with a higher yield stress can be used.

With steel, we are usually able to select filler metal having suitable mechanical properties, i.e. we can overmatch the parent metal. With many non-ferrous metals and alloys this is often not the case. In the first place, these materials tend to be weaker if they have a cast structure − in the weld for example. It may also be necessary to use a filler-metal composition which gives a lower strength in order to avoid or minimise cracking in the weld metal. Undermatching is therefore commonplace with non-ferrous metals, and, when a tensile load is applied, yielding first takes place in the weld metal. Fracture occurs in the weld, and the tensile strength of the joint is below that of the parent metal. The extent of this difference can be expressed in terms of joint efficiency:

$$\text{joint efficiency} = \frac{\text{tensile strength of joint}}{\text{tensile strength of parent metal}} \times 100\%$$

Summarising, in the case where yielding in the tensile test occurs away from the weld, as with overmatching, the allowable design stress is that of the parent metal. On the other hand, with joint efficiencies of less than 100% the margin between the parent-metal design stress and the stress at which yielding occurs is reduced, and some adjustment to working-stress levels must be made.

## 7.3 Strength of fillet-welded joints

The fillet-welded joint requires a very different approach because, unlike the butt weld, the size of the fillet is controlled by the welder and the throat dimension can be increased at will. In a 'T' joint, the throat is taken as the shortest distance from the surface of the weld to the intersection of the weld and the joint at the root (fig. 7.2). This includes the excess metal and, as with the butt weld, it is customary for design purposes to neglect the contribution this part of the weld makes to strength. We must, therefore, redefine the design throat as the shortest distance from the root to the line drawn between the toes of the weld. It is assumed that fracture under load will occur across the throat. In practice, weld-metal failures in fillet-welded joints under test often tend to follow a line closer to the fusion boundary, but the error involved in using the design throat as defined above is relatively small. The point is somewhat academic, since our aim is to design a joint which will *not* fail under the loads experienced in service.

The load-carrying area of the fillet weld is the design-throat thickness multiplied by the length of the weld. In fillet welds which transmit loads, the throat area is in shear and the maximum allowable shear stress will be determined by the composition of the weld deposit, which in turn depends on the type of electrode used and the material being welded. As an example of this, the information for fillet welds in Table 7.1 is taken from British

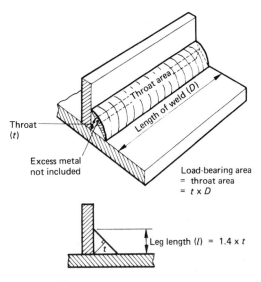

Throat
(*t*)

Excess metal
not included

Throat area

Length of weld (*D*)

Load-bearing area
= throat area
= *t* x *D*

Leg length (*l*) = 1.4 x *t*

**Fig. 7.2** Throat thickness and leg length in fillet welds

Standard BS 2573:part 1:1983, which is concerned with crane jibs made
from carbon–manganese steels to BS 4360 grade 43 and grade 50.

**Table 7.1** Strength of fillet welds

| BS 4360 grade | Plate properties (N/mm²) | | Max. allowable shear stress on fillet welds (N/mm²) | |
|---|---|---|---|---|
| | Tensile strength | Min. yield stress | Electrode E43 | Electrode E51 |
| 43 | 430–510 | 245 | 118 | 126 |
| 50 | 500–620 | 355 | 118 | 144 |

The calculation of the size of fillet for a given application is based on the
concept that the throat area should be large enough to transmit the maximum
tensile stress allowed for the parent metal. The length of the weld is usually
dictated by the dimensions of the component or structure, and the calculation
aims to specify the minimum throat thickness. This dimension is of little
direct help to the welder, since it is difficult to estimate the throat thickness
while depositing the weld, especially if it contains more than one run. It is
more normal to quote the leg length in instructions given to the welder.

### 7.4 Effect of heat-affected-zone properties
In chapters 2 and 3, the flow of heat from the weld pool into the parent
metal was discussed as an essential part of the solidification processs. A

secondary effect of the heat flow is to raise the temperature of the parent metal adjacent to the fusion boundary. A number of metallurgical reactions can take place in this region. They affect the mechanical properties of the resultant joint either by causing a loss of tensile and impact strength, an increase in hardness, or the formation of cracks.

The structure of the parent metal which has been raised in temperature depends on its composition and on the heating and cooling cycles (fig. 7.3). The boundary of this heat-affected zone (h.a.z.) is defined by the lowest temperature at which metallurgical change takes place.

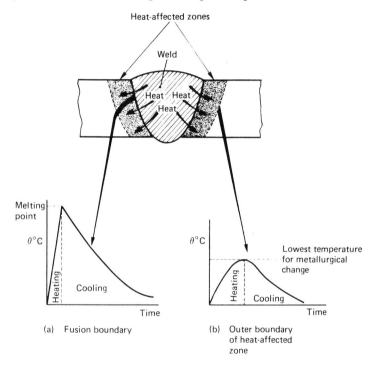

Fig. 7.3    Heat-affected-zone boundaries

**Work-hardened metals** are characterised by increased hardness and tensile strength, resulting from deformation, or working, at room temperature. Sheet material is often delivered in the word-hardened condition since the manufacturer has rolled it cold to give acceptable surface finish, accurate thickness, and higher strength. We also meet with work-hardening in fabrication, where sheet is shaped by pressing, bending, or forming at room temperature.

The effects of this cold working can be removed by heating to a critical temperature above which softening occurs. This is known as recrystallisation. During welding, the h.a.z. experiences temperatures above this critical

102

temperature. The h.a.z. is therefore softened, and its tensile strength falls below that of the unaffected parent metal, which still has the work-hardened properties (fig. 7.4). It is not practicable to recover hardness in the h.a.z. since this would require extensive cold rolling of the joint, associated with a significant reduction in thickness. As the strength of the joint has been permanently reduced by softening in the h.a.z., the maximum allowable stress is lower than that for unwelded metal.

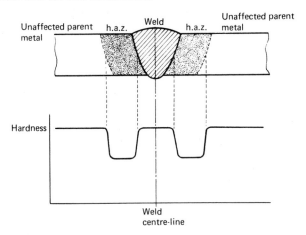

**Fig. 7.4**   Hardness variations in a welded joint in work-hardened material

**Precipitation-hardened alloys** derive their strength from a controlled heat treatment which is often called age-hardening. The heating cycles in the h.a.z. continue the heat treatment beyond the optimum time at a temperature leading to an over-ageing effect which softens the metal and results in a loss of strength. In addition, the weld metal will frequently have low strength. By contrast with work-hardened metals, the properties of precipitation-hardened alloys can be recovered by heat-treating the completed joint. This may not always be feasible, and in many instances the component must be used in the as-welded condition, accepting an appreciably lower permissible stress. For example, with an aluminium alloy containing 4% copper, the permissible axial stress for the heat-treated plate is 131 $N/mm^2$ whereas for the as-welded joint the corresponding figure is 51 $N/mm^2$.

**Structural steels** tend to harden in the h.a.z. because the cooling rates encountered in welded joints give quenching effects similar to those used in heat-treatment practice. The amount of hardening depends, in the first place, on the composition of the steel. Structural steels contain a number of alloying elements, and for convenience in discussing the response to welding it is common practice to use an index known as the carbon equivalent (c.e.). In this, the amount of each alloying element present is factored according to the contribution it makes to hardening. Thus, manganese is allocated a factor of one sixth, since it has been estimated that a manganese addition of 0.6% produces the same effect on quenching as 0.1% carbon. There are a number of formulae used to determine the carbon equivalent; British Standards quote

$$c.e. = C\% + \frac{Mn\%}{6} + \frac{(Cr + Mo + V)\%}{5} + \frac{(Ni + Cu)\%}{15}$$

where  C = carbon       Mn = manganese      Cr = chromium
       Mo = molybdenum     V = vanadium        Ni = nickel
and    Cu = copper

In general terms, steels with c.e. values less than 0.38% experience very little hardening in the h.a.z., whereas at a c.e. of 0.50% the cooling cycles associated with welding can lead to a hardness comparable with those produced in tools by heat treatment. On the Vicker's diamond scale these would be about 400 VPN (Vicker's Pyramid Number), compared with the 190 VPN normally measured on low-carbon steel. Usually the fact that the h.a.z. is hard does not affect the strength or service suitability, although occasionally special considerations such as stress corrosion attack may require a control of the maximum hardness, and this could involve heat treatment after welding.

The main concern arising from hardening in the heat-affected zone is the formation of cracks, which will drastically reduce the strength of the joint. If hydrogen is present in solution in the steel alongside the fusion boundary, there is a high risk of cracks being formed when the hardness exceeds about 300 VPN (fig. 7.5). Such a hardness level is associated with metallurgical changes in the h.a.z. and is controlled by composition and cooling rate. This poses a major problem in welding, since higher strength steels almost invariably have a higher c.e. than low-carbon steel. When steels with c.e.'s of 0.40% and above are welded, procedures are required which not only reduce the cooling rate but also keep the hydrogen content to a minimum.

Hydrogen in the weld pool originates from a number of sources, such as the electrode coating, moisture in the gas, and grease on the filler wire and plate. It is dissolved by the molten metal and flows into the parent metal, where it concentrates in the h.a.z. Hydrogen contents of weld pools have been measured, and critical values have been defined for given hardness levels. These contents are expressed in millilitres of hydrogen per 100 grams of weld metal (see page 80) and usually range from 5 to 30 ml/100 g.

The actual distribution of hydrogen in a welded joint depends to some extent on the temperature of the parent metal. Hydrogen diffuses quickly through hot metal but moves very slowly at room temperature; thus it will readily flow into the h.a.z. but will not move outside the heated band during the welding cycle. Preheating the plate enables the hydrogen to disperse over a wider area, so reducing its concentration in the hardened region. At the same time, preheating slows the cooling rate and lowers the maximum hardness. For these reasons, preheat temperatures between 75 and 200°C are commonly used when welding higher strength structural steels.

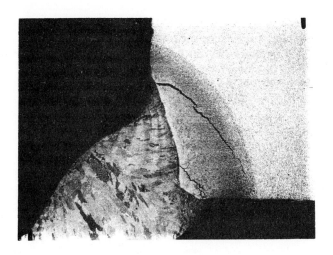

**Fig. 7.5** H.A.Z. cracks in a high-yield-strength steel

Data are available in documents such as BS 5135:1974, 'Metal arc welding of carbon and carbon–manganese steels', enabling the welding engineer to specify a weld procedure which will avoid h.a.z. cracking. A number of factors need to be considered and a choice, based on experience, must often be made from three or four possible combinations of welding variables. Some of the important aspects are summarised below.

***Summary of factors governing h.a.z. cracking in welded joints in structural steels***
a) Hardness in the h.a.z. depends on the cooling rate (fig. 7.6) and the c.e.

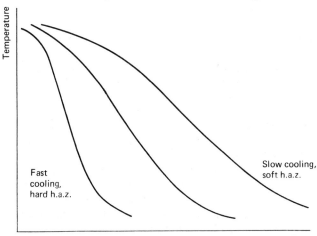

**Fig. 7.6**

105

b) If the h.a.z. hardness exceeds a critical level, there is a risk of cracking.
c) The hardness which can be tolerated depends on the hydrogen content of the weld. This means that with low hydrogen contents higher hardnesses – i.e. faster cooling – can be tolerated.
d) For each combination of c.e. and hydrogen level there is a cooling rate which gives the critical level of hardness (figs 7.7 and 7.8).

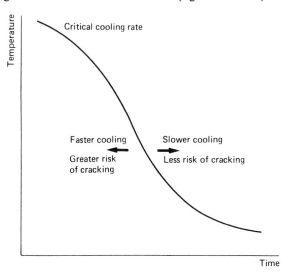

**Fig. 7.7**

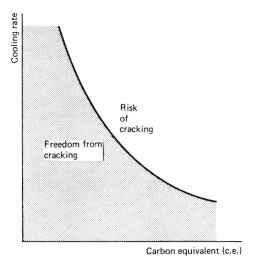

**Fig. 7.8**   Combinations of cooling rate and c.e. within the shaded area are unlikely to lead to cracking in the h.a.z.

e) In practice, the cooling rate is controlled by heat input, preheat, and thickness (fig 7.9). The cooling rate is reduced by increase in heat input, increase in preheat, and reduction in thickness. Note: preheat also helps to disperse hydrogen from the h.a.z.

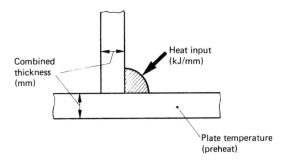

**Fig. 7.9**

## 7.5 Influence of weld defects

We can now return to the subject of defects and the influence they have on the service performance of welded joints. There is a host of possible defects, which have been classified in various documents such as British Standard BS 499:part 1:1983. For simplicity we will select six typical and very common defects (fig. 7.10): undercut, cracks, porosity, slag inclusions, lack of fusion, and lack of penetration.

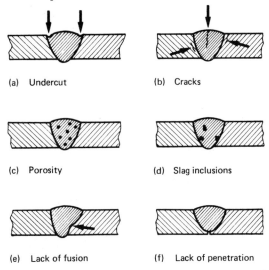

(a) Undercut          (b) Cracks

(c) Porosity          (d) Slag inclusions

(e) Lack of fusion    (f) Lack of penetration

**Fig. 7.10** Cross-sections of welds containing typical defects

107

The first thing one notices is that in a weld cross-section these defects provide discontinuities in the path of a transverse tensile stress, and it would be tempting to suggest that their significance is in the reduction of the effective load-bearing area. In fact, in many cases a reduction of up to 28% in cross-sectional area could be tolerated before yielding would take place in the weld rather than the plate, and this reduction would need to be along the complete length of the weld. This might occur in the case of incomplete penetration, but it is most unlikely that undercut, porosity, cracks, slag inclusions, or lack of fusion could produce such a significant loss of cross-section. We must therefore look to another reason why we do not accept defects.

### Stress concentrations

Consider a plate subjected to a tensile load (fig. 7.11). The stress is uniformly distributed across the plate and is given by

$$\sigma_1 = \frac{\text{load } (F)}{\text{area } (A)} = \frac{F}{\text{width } (w) \times \text{thickness } (t)}$$

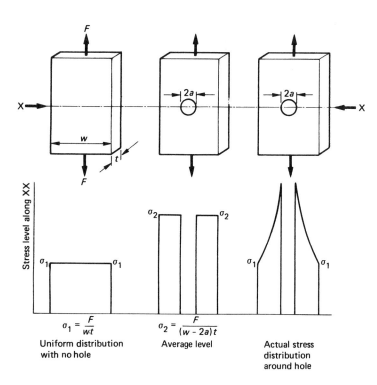

Fig. 7.11  Stress concentration around a hole in a bar

108

If a hole of radius $a$ is drilled at the centre of the plate, the width of the plate is reduced by $2a$. The load is therefore being carried by a smaller cross-section and the stress rises to

$$\sigma_2 = \frac{F}{(w - 2a) \times t}$$

This is the mean stress level; the actual stress distribution is not uniform but is high at the edges of the hole, falling to the original level $\sigma_1$ some distance away.

The discontinuity in the path of the stress flow created by the hole has caused a concentration of stress. The severity of this stress-concentration effect depends on the shape of the discontinuity: with a round smooth hole the concentration factor is low, being of the order of 3, while a sharp notch can give stress concentrations as high as 10. We can see, therefore, that the real significance of defects is their ability to create stress concentrations. This means that we can be more tolerant of rounded discontinuities such as gas pores and slag inclusions than of defects such as cracks. Practical confirmation of this view is to be found in various standards; Table 7.2 gives the allowances for defects specified in BS 5500, which is concerned with pressure vessels.

**Table 7.2** BS 5500 allowance for defects

| Defect type | Acceptance criteria |
|---|---|
| Porosity – isolated | Diameter can be between 3 mm and 6 mm, depending on plate thickness. |
| Porosity – distributed | 2% of the area of the weld metal as seen on a radiograph. |
| Slag inclusions | Length can be numerically equal to plate thickness but not greater than 100 mm. |
| Incomplete root penetration | Not permitted. |
| Lack of fusion | Not permitted. |
| Cracks (any type) | Not permitted. |

### Notch ductility

In making a full assessment of the significance of a defect, reference must be made to the material in which it exists. A crack transverse to the direction of applied tensile stress (fig. 7.12) creates a high concentration at its ends.

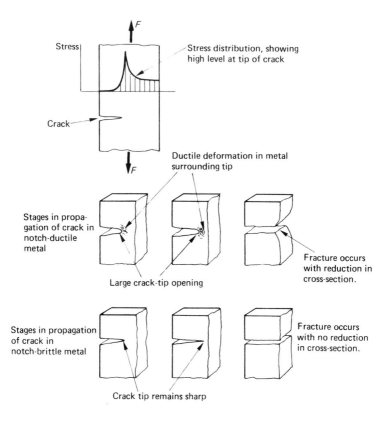

**Fig. 7.12** Comparison of notch-ductile and notch-brittle behaviour

In ductile material, yielding takes place in the material surrounding the tip of the crack, and the width of the crack increases at the same time as the crack extends in length. The continued behaviour of the crack depends on a number of factors, and in certain circumstances the crack may not propagate. If it does proceed across the complete width of the component, the final fracture shows evidence of appreciable plastic deformation. On the other hand, in some materials there is very little opening of the crack, which then propagates very rapidly across the specimen, giving a fracture in which there has been virtually no reduction of area, i.e. a brittle failure. Such a material is said to be notch brittle, as opposed to the notch-ductile case quoted before; alternatively, it can be described as having low notch ductility.

The problem of notch-brittle behaviour is a very real consideration in the design of structures fabricated in carbon or carbon–manganese steels, since these materials can exhibit both notch-ductile and notch-brittle behaviour, depending on temperature. A series of Charpy tests (fig. 7.13) performed at

110

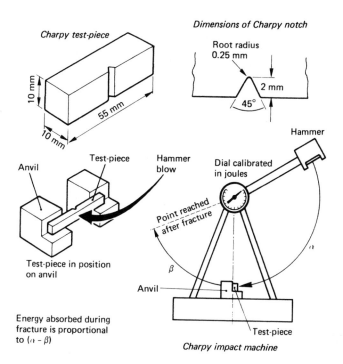

**Fig. 7.13** Charpy impact testing

different temperatures ranging from, say, −60 to +60°C on specimens cut from steel plate gives the sort of results shown graphically in fig. 7.14. Although there is usually considerable scatter, a clear trend can be observed from a high level of energy absorbed during fracture with a marked fall to very low energy values.

The temperature at which this transition takes place is a characteristic of the composition of the steel, its grain size, and the treatments it has received. Below the transition temperature a planar defect such as a crack or lack of fusion behaves in a notch-brittle manner and a fast-running brittle fracture could occur. It is because of risks of this type that standards such as BS 5500 are severe in their treatment of planar defects, while other standards concerned with different materials or less stringent situations are more prepared to accept, say, lack of fusion. In general, cracks of significant size are never tolerated.

*Establishing standards of acceptance*
With very few exceptions, the acceptance standards quoted and used by different authorities reflect years of experience and are aimed at a 'safe' situation. Inevitably this sometimes means that the design may not make

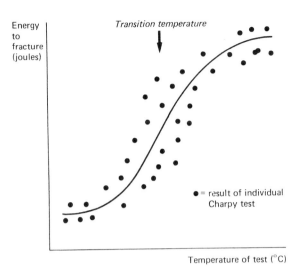

Fig. 7.14 Effect of test temperature on results of Charpy tests of carbon-steel specimens

best or most efficient use of material, but often this is a small price to pay for confidence in the ability of the structure to withstand service requirements.

A problem which confronts anyone who sets out to define an acceptance standard is the imprecise nature of the measurements. In general engineering it is relatively straightforward to stipulate that the diameter of a shaft should be 100 mm ± 0.1 mm, since the inspector can readily measure this and reject shafts with diameters which are more than 100.1 mm or less than 99.9 mm. On the other hand, how does one specify the maximum amount of porosity that can be allowed in a joint? We could say 'not more than 1 pore per 10 mm of weld', which seems explicit enough, but on reflection there are a number of questions we could ask. What diameter of pore would we tolerate? Is this an average? Would three pores in the middle 10 mm of a 30 mm length be acceptable? Similarly, how do we measure lack of fusion, which will have depth as well as length in a weld?

There is no easy answer to these problems and much depends on the experience both of the writer of the standard and of the inspector or surveyor who is interpreting the acceptance criteria. There are many examples of good approaches to be found in standards such as BS 5500, BS 2633 and ASME 9, and these are worth studying.

A more precise approach is offered by the rapidly developing technique of fracture mechanics. If we know the size and location of a planar defect and the stress surrounding it, and we have some measure of the notch ductility of the material, it is possible to predict whether the defect will grow and lead to failure in service. To do this requires a quantitative representation of the notch ductility. Unfortunately, Charpy tests do not give data which can be used in calculations. To get some idea of what we need, we can look back to fig. 7.12 and notice that the difference between ductile and brittle behaviour is typified by the amount of deformation at the tip of the crack. This can be quantified in a CTOD (crack tip opening displacement) test to give data for use in fracture calculations.

In this test (fig. 7.15), a slot is cut in the sample and a crack is formed at its root by subjecting the sample to fatigue loading. The sample is cooled to the test temperature and then bent around a support positioned below the crack. Stress is developed at the tip of the crack, and the material there will either deform plastically or fracture with no loss in section. A clip gauge placed at the mouth of the slot measures the amount of displacement.

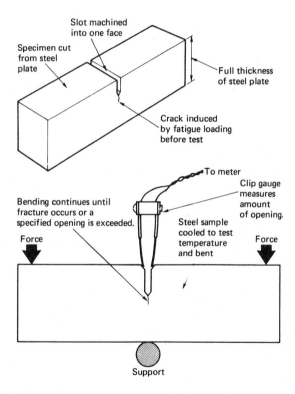

**Fig. 7.15** CTOD test

# 8 Checking and controlling weld quality

## 8.1 Non-destructive testing

When weld quality was being discussed in chapter 7, we recognised the importance of some defects in terms of service behaviour and crack propagation. It follows that if we are to use welded structures with confidence we must have some means of checking whether unacceptable defects are present in the welds. It is also clear that, after testing, the welded joint must be in a fit condition for use. In other words we must use non-destructive testing (NDT) methods.

There are many NDT techniques available to the weld inspector, but they are mostly expensive either in terms of capital equipment or because labour costs are high. Each method has its own limitations and there is no universal testing system. The inspection of welded joints must be carefully planned to ensure that the selected technique is capable of detecting the defects which are of concern to us.

It is also important to specify the acceptance standard *before* testing is undertaken, since this will dictate the sensitivity required. We can take sensitivity in this context as meaning the smallest defect which can be located with certainty. There is no point in stating that cracks with lengths greater than 1 mm are unacceptable if the test method cannot detect cracks less than 2 mm long – such a situation simply gives a false sense of security. It would also be wrong to say that a weld is free of defects because the inspection technique used did not discover any. A more sensible statement would be that there were no flaws detected.

We must also recognise that NDT of itself cannot tell us if a weld is acceptable or not. All we can report is a list of the types of defects discovered, their size, and their location in the joint. It is for the accepting authority, be it the insurance surveyor, customer's representative, or in-house inspector, to decide if the level of defects is greater than that which the designer would be prepared to tolerate in the service environment of the fabrication.

## 8.2 Visual inspection

It is a common mistake to think that NDT as applied to welded joints consists exclusively of radiography or ultrasonic flaw detection, which we shall discuss later. Before either of these techniques is used, the experienced inspector will examine the joint visually. Often defects can be discovered by the naked eye and can be rectified or repaired at this stage.

In visual inspection, attention is paid to three aspects:

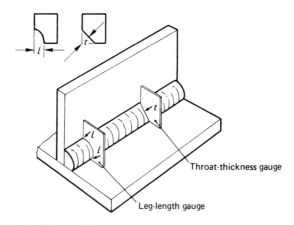

**Fig. 8.1**    Simple fillet-weld gauges

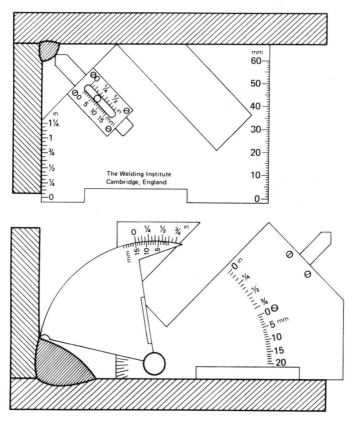

**Fig. 8.2**  Universal weld gauge (Welding Institute): *top* throat measurement; *bottom* leg-length measurement

115

a) dimensions of the weld (especially in fillet welds, figs 8.1 and 8.2),
b) penetration in joints welded from one side,
c) surface defects.

## 8.3 Assisted visual inspection

Even with good lighting, it is often difficult to detect cracks by visual inspection. The effectiveness and indeed the speed of the examination is improved if the presence of a crack is highlighted by the use of either dye-penetrant or magnetic-particle techniques.

### Dye-penetrant inspection

If a liquid which has a low surface tension is poured on to the surface of the welded joint, it wets the metal and flows uniformly (fig. 8.3). In particular, it seeps into a crack or cavity. Wiping the surface of the parent metal and weld leaves the liquid in the crack. If an absorbent layer of chalk is now deposited on the clean surface, the liquid is drawn out of the crack. By adding a red dye to the liquid, the location of the crack is indicated by a red stain on the chalk. Alternatively, the liquid may contain a dye which fluoresces when viewed under ultra-violet light.

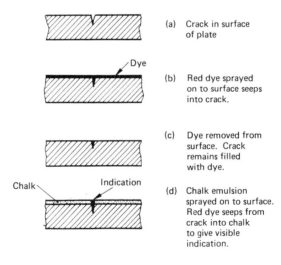

(a)  Crack in surface of plate

(b)  Red dye sprayed on to surface seeps into crack.

(c)  Dye removed from surface. Crack remains filled with dye.

(d)  Chalk emulsion sprayed on to surface. Red dye seeps from crack into chalk to give visible indication.

**Fig. 8.3**   Principles of dye-penetrant inspection

Dye-penetrant inspection is very easy to use on the shop floor or on site, since the equipment required is very portable. The dye, dye remover, and chalk 'developer' are obtainable in aerosol cans. The choice between chalk developer and fluorescent light is very much a matter of judgement based on local circumstances. Fluorescent dye may give somewhat better definition of the outline of the crack, but it is necessary to view in subdued background light. In addition, the ultra-violet illuminator requires a power supply. Against this, the use of a chalk developer requires more time and, if left, the dye spreads across the surface of the chalk, making the indication more diffuse.

### Magnetic-particle inspection

When inspecting ferromagnetic materials such as carbon steel, an alternative method of locating surface cracks is magnetic-particle inspection (MPI). In this technique, a magnetic field is created in the crack and is used to attract iron-oxide particles.

Consider a magnet in the shape of a bar (fig. 8.4). The lines of force within the magnet run from one end of the bar to the other, i.e. from the S to N poles. At the same time a magnetic field exists around the magnet. Iron powder sprinkled on to the bar collects around the poles, where the lines of force are close together. If another magnet is brought into proximity so that its N pole is opposite the S pole of the first magnet, the lines of force flow between the two magnets. When iron powder is sprinkled on to the new arrangement it collects in and around the gap between the N and S poles.

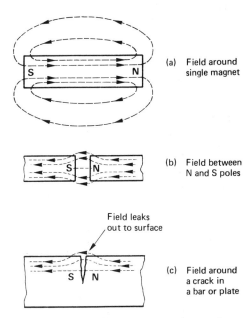

(a) Field around single magnet

(b) Field between N and S poles

Field leaks out to surface

(c) Field around a crack in a bar or plate

**Fig. 8.4** Magnetic fields around bar magnets and a crack

We can make a crack in a weld in steel behave in the same way as the gap between the magnets by inducing a magnetic field at right angles to it. The lines of force flow across the crack and 'leak' out to the surface. Iron or iron-oxide particles deposited on the weld collect at the point where the magnetic field is leaking, thus indicating the presence of a crack.

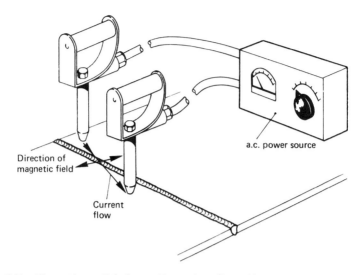

**Fig. 8.5** Magnetic-particle inspection using alternating current

Permanent or electromagnets can be used to create a magnetic field in the welded joint. More commonly, a current is passed through the area being examined (fig. 8.5). The detecting fluid or 'ink' is usually particles of magnetic iron oxide suspended in kerosene – the particles may be coated with a fluorescent compound for viewing under ultra-violet light. In cases where there are objections to the use of kerosene, the magnetic particles can be dispersed from a 'puffer' to form a cloud above the surface of the weld. This alternative is particularly useful when inspecting root runs in 'V' joints, since there is no need to clean the groove before depositing the next run of weld metal.

## 8.4 Radiography

Visual inspection, with or without the aid of dye-penetrant and magnetic-particle techniques, can give positive information only about defects which appear on the surface. Traditionally, radiography has been used to locate defects in the body of the weld, although in recent years ultrasonics has also become recognised as an established method of non-destructive testing.

The detection of defects in welds by radiography is based on the ability of X-rays and gamma rays to penetrate materials which are opaque to ordinary white light. Both are electromagnetic rays and have characteristic wavelengths:

X-rays $2 \times 10^{-12}$m to $10^{-9}$m
gamma-rays $10^{-13}$m to $2 \times 10^{-12}$m

X-rays are produced at will in a suitable vacuum tube, while gamma rays are emitted by radioactive material. In terms of flaw detection they behave in similar ways, and for the following explanation of the principles of radiography we will not distinguish between them. Later in the chapter their specific characteristics will be discussed.

During transmission through the material the X-rays are absorbed. If the material is homogeneous, the amount of absorption is uniform across the area exposed to the X-ray beam. On the other hand, if the material contains, say, a pore of gas, a smaller amount of the rays passing through this point is absorbed and there is a variation in the intensity of the emergent beam. This can be readily detected by placing a photographic film on the side of the material opposite to the source of the radiation (fig. 8.6). The film is exposed by the X-rays and its optical density depends on the intensity of the radiation. Thus the area under the gas pore receives more X-rays than the parent material on either side and the photographic emulsion is affected to a greater extent. On a negative film this shows up as a dark spot of the same shape as the pore (fig. 8.7).

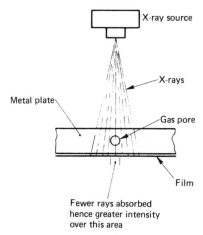

**Fig. 8.6**  Principles of radiography

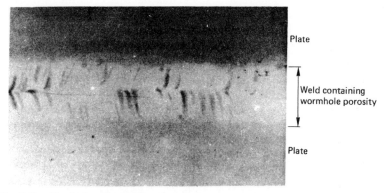

**Fig. 8.7**  Typical radiograph

The difference in intensities of the emergent rays depends on the relative absorption coefficients of the parent material and of the gas in the pore. The coefficient of absorption is a function of the atomic number of the material and the wavelength of the X-rays. Fortunately, defects in welded joints usually contain either gas, air, or slag, which have appreciably smaller absorption coefficients than the parent metal. Their presence is thus readily detected, since there is a marked variation in density in the film. The only exception to this is provided by tungsten inclusions: this metal is denser than those normally welded and gives a bright spot on the negative film, since more X-rays are absorbed.

Not all defects are detected by radiography – much depends on the orientation of the flaw with respect to the beam (fig. 8.8).

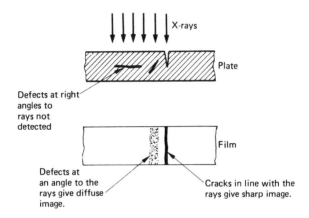

**Fig. 8.8** Orientation of defect with respect to X-rays

*Taking a radiograph*
When a weld is to be radiographed, any extraneous matter on the surfaces is first cleaned off to ensure that there are no spurious indications. Sometimes the excess metal and underbead are removed by grinding, to give more uniform conditions.

The film is held in a lightproof cassette which is placed against one side of the joint. The X-ray source is fixed at a suitable distance from the weld on the opposite side of the joint and is arranged so that the beam falls symmetrically on to the surface of the metal. The source is switched on and the joint is exposed to the X-rays for a predetermined period. Exposure times depend on the intensity of the radiation, the wavelength of the rays, and the thickness of the metal being examined – they can range from 10 seconds to 10 minutes for X-rays and 1 minute to 24 hours for gamma rays. After exposure, the cassette is taken to a dark-room where the film is developed and fixed in a similar way to normal photographic film. When it has dried, the film is viewed against a bright light.

One of the advantages of radiography is that the film can be stored as a permanent record of the quality of the weld.

### Generation of X-rays

X-rays are generated in a vacuum tube which contains a cathode and an anode (fig. 8.9). The heated cathode emits electrons which are accelerated towards the anode by a high voltage difference of the order of 100 to 400 kV. The quantity of electrons depends on the temperature to which the cathode is heated and is therefore a function of the current flowing in the cathode filament (2 to 10 mA). The anode consists of a tungsten disc (called the anticathode or target) set in a copper block. As the electrons hit the target, their kinetic energy is released with the generation of both X-rays and heat. The heat is conducted away by the copper block, which in high-power tubes may be liquid-cooled. The target is at an angle to the beam of electrons, so that the X-rays are emitted through a window towards the work.

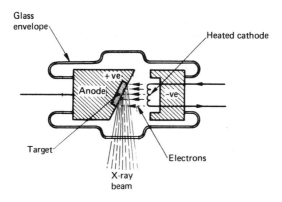

**Fig. 8.9**   Simplified construction of an X-ray tube

### Gamma-ray sources

Some elements have unstable atomic nuclei which disintegrate with the passage of time. This disintegration is accompanied by the emission of radiation, part of which is composed of gamma rays. These have high energy and can penetrate metals in the same way as X-rays. They can therefore be used for the examination of welded joints.

For the inspection of welds, 'artificial' radioactive isotopes, which are obtained by fission or irradiation in a nuclear reactor, are used to provide a source of gamma rays. The most popular isotopes are iridium-192, cobalt-60, caesium-137, and caesium-134.

Isotopes are small and, unlike X-ray sources, do not need a power-supply unit. They are readily portable, but must be handled with care and stored in lead-lined containers. The gamma rays are radiated in all directions, a feature which can be of particular value when inspecting pipe welds, as the source can be located in the bore with the film wrapped around the outside of the pipe.

## Safety in the use of radiography

X-rays and gamma rays can be extremely harmful to human beings since they damage body cells. Depending on the dosage received, exposure to this form of radiation can result in ailments ranging from burns to leukemia and cancerous growths. The industrial use of radiography is the subject of strict legalislation (e.g. *Ionising radiation regulations*, HMSO) designed to protect not only the people taking the radiograph but also those engaged in other operations in the vicinity.

The effects of exposure to radiation are cumulative, and persons working continuously with radiographic equipment are required to wear film badges which measure the amount of radiation they have received over a stated period of time. This is known as the dose rate and, in practice, provisions must be made to keep it below a certain level. In addition, the testing area should be screened to ensure that the radiation does not reach other workers. The screens are preferably lead-lined or are made of lead-bearing brick.

X-rays are usually more manageable since they are highly directional – whenever possible, the tube and work should be arranged so that the beam is pointing at the floor. Gamma rays are given off in all directions from an isotope and pose greater problems in screening. The area in which radiation may exist must be identified by the use of the international warning symbol. After the film has been exposed, an X-ray machine can be switched off and is then safe. On the other hand, a radioactive isotope emits radiation continuously and must be kept in a shielded container when not in use.

## 8.5 Ultrasonic inspection

Ultrasonic inspection offers a highly sensitive method of detecting flaws inside a metal by observing the way in which high-frequency vibrations are transmitted through the metal. If the material being tested contains a flaw, the vibrations are reflected. The presence of the flaw can be indicated either by noting the reduction in the strength of the transmitted vibrations or by · monitoring the reflections.

This type of behaviour is typical of a wide range of vibration frequencies. We are familiar with the reflection of sound waves from a wall or a sheet of metal; this happens at frequencies from 16 Hz, corresponding to the lowest note on an organ, up to the point at which we can no longer hear the sound, somewhere between 15 and 20 kHz depending on the individual person.

## Detecting ultrasonic waves

Reflections from an interface still occur at higher frequencies, i.e. in the ultrasonic range, but we cannot recognise the effect with the unaided ear. Instead, we must use some kind of sensor. In the ultrasonic testing of metals, frequencies between 0.5 and 10 MHz are used, and the vibrations are measured by using the piezo-electrical characteristics of a crystal.

If a disc of quartz of uniform thickness is subjected to a pressure, positive and negative charges are produced on opposite faces. When the pressure is removed and a tensile force is applied through the thickness, the charges are reversed. By placing the quartz crystal in the path of the ultrasonic vibrations it is alternately subjected to compression and tension, and the charges on its face alternate between positive and negative. These charges can be fed to a cathode-ray oscilloscope where they can be displayed, thus giving a useful method of detecting the presence of ultrasonic vibrations (fig. 8.10).

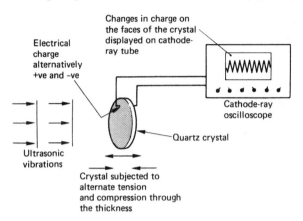

**Fig. 8.10** Using a quartz crystal to detect ultrasonic vibrations

*Generation of ultrasonic vibrations*
We can also use the quartz crystal as a generator of vibrations by reversing the process. If an alternating electric potential is applied to it, the crystal emits ultrasonic vibrations. A problem arises when we try to inject these vibrations into the piece of metal being tested. Simply placing the crystal in contact with the metal surface leaves an air gap, giving an interface which reflects or attenuates (i.e. reduces) the signal. Coating the surface with oil or grease overcomes this by ensuring good contact between the quartz and the metal.

*Ultrasonic testing of welded joints*
There are two ways of using ultrasonic waves in the testing of welded joints: transmission and reflection.

**Transmission** Separate transmit and receive probes are placed on opposite sides of the plate. If there are no defects then there is only a slight attenuation of the signal which reaches the receiver (fig. 8.11). A small defect in the path of the ultrasonic beam reflects some of the waves, and the signal detected on the receiver side is reduced. On the other hand, a large defect such as a lamination in the plate causes a complete loss of the transmission signal.

With this method of testing there is no way of determining how deep the defect is below the surface. Major drawbacks are the need to have good access

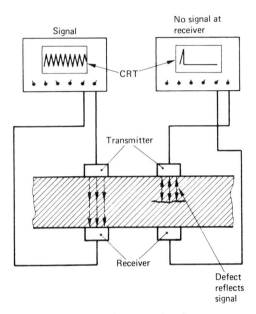

**Fig. 8.11** Transmission technique for detecting flaws

to both sides of the component or section and the ever-present problem of ensuring accurate alignment of the probes. If the probes are not directly opposite each other, the indication will be as if there were a defect present.

**Reflection**　A transmitter probe is used for this method and access is required to only one surface.

With a *normal probe*, the signal is transmitted at right angles to the surface (fig. 8.12). In the absence of a defect, the ultrasonic waves travel through the thickness of the plate, are reflected from the other surface, and return to the probe. They are displayed on the cathode-ray tube (CRT) as a boundary echo. The CRT also records the entry of the waves into the plate as the transmitter signal. The distance between this and the boundary echo is a measure of the thickness of the plate.

When there is a defect in the path of the waves, some of the signal is reflected and, having a shorter distance to travel, appears on the CRT before the boundary echo. The depth of the defect below the surface can be deduced from its position relative to the transmitter and boundary indications. Separate transmit and receive crystals mounted in one probe can be used, but a single crystal transmitting and receiving pulses is probably more common in current practice.

An *angle probe* fitted with a single crystal is normally used to test welded joints (fig. 8.13). A pulse of ultrasonic vibrations is transmitted into the plate at an angle to the surface. This travels to the opposite surface, where it is reflected at an angle equal to the angle of incidence. The vibrations are again

124

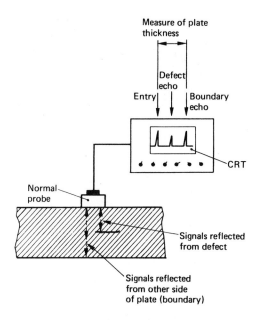

**Fig. 8.12** Reflection technique using a normal probe which injects pulses of ultrasonic vibrations at right angles to the plate surface

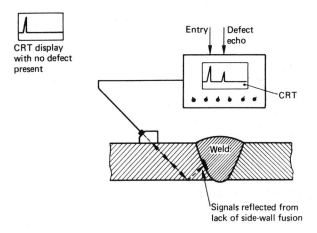

**Fig. 8.13** Reflection technique with an angle probe

reflected from the top surface, and so on. If there is no interface in their path, the vibrations continue until they peter out. On the other hand, a defect reflects some or all of the signal, which may then follow the original path or take a new route back to the probe, depending on the orientation of the

125

interface with respect to the beam. A defect echo is displayed in the CRT, and again its distance from the transmitter signal will be a measure of the distance travelled by the ultrasonic waves. Knowing the geometry of the system, the exact location of the defect can be calculated.

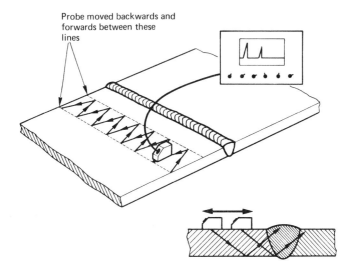

**Fig. 8.14** Moving the probe to scan the thickness of the weld

In the examination of a weld (fig. 8.14), the operator marks two lines parallel to the weld and moves the probe backwards and forwards between these two limits to ensure that the complete thickness of the weld is examined.

Table 8.1 summarises the methods of non-destructively testing welds.

### 8.6 Controlling weld quality
The non-destructive tests described in sections 8.2 to 8.5 are of considerable value since, by providing details of any defects which can be located, they give us a degree of confidence in the ability of the welded joint to withstand service conditions. This is all they can do. As we have already discussed, they will not tell us if the weld is good enough for our purposes; equally important, they will not guarantee that welds will be made which are free of defects. In other words, we cannot control quality simply by specifying that the welds will be subjected to NDT on completion. Admittedly the knowledge that the welds will be tested may act as an incentive to do better work, but it is far more relevant to establish a system which encourages the production of defect-free welds at all times.

Welding is usually the last of the series of operations involved in fabricating a structure, and the quality of the weld reflects what has gone before.

126

**Table 8.1**   Summary of the methods of non-destructively testing welds

| Method | Defects detected | Advantages | Limitations |
|---|---|---|---|
| Visual | Inaccuracies in size and shape. Surface cracks and porosity, under-cut, overlap, crater faults. | Easy to apply at any stage of fabrication and welding. Low cost both in capital and labour. | Does not provide a permanent record. Provides positive information only for surface defects. |
| Dye-penetrant | Surface cracks which may be missed by naked eye. | Easy to use. No equipment required. Low cost both in materials and labour. | Only surface cracks detected with certainty. No permanent record. |
| Magnetic-particle | Surface cracks which may be missed by naked eye. May give indication of sub-surface flaws. | Relatively low cost. Portable. Gives clear indication. | Only surface cracks detected with certainty. Can be used only on ferro-magnetic metals. Can give spurious indications. No permanent record. |
| Radiography | Porosity, slag inclusions, cavities, and lack of penetration. Cracks and lack of fusion if correctly orientated with respect to beam. | Can be controlled to give reproducible results. Gives permanent record | Expensive equip-ment. Strict safety precautions required. Better suited to butt joints — not very satisfactory with fillet-welded joints. Requires high level of skill in choosing conditions and interpreting results. |
| Ultrasonics | All sub-surface defects. Laminations. | Very sensitive — can detect defects too small to be discovered by other methods. Equipment is portable. Access required to only one side. | Permanent record is difficult to obtain. Requires high level of skill in interpreting cathode-ray-tube indications. |

This point is often overlooked when considering the control of weld quality, and welders can justifiably claim that on many occasions they are blamed for faults which are the direct result of inadequate preparation for welding. We must take an overall view of the manufacturing operation and at each stage look for factors which could subsequently influence the formation of weld defects.

In fabrication work, *quality assurance* is concerned with defining and planning *all* the operations necessary to manufacture a product which will perform satisfactorily in service but which has been completed in the most economical manner. It covers design, evaluation of the manufacturing procedures, and definition of inspection criteria for each stage, including testing the final product. We can differentiate between this and *quality control* because the latter relates to the system established to monitor each stage of manufacture with a view to achieving the requirement specified by quality assurance.

### 8.7 Approving the weld procedure and the welder

The procedure used to deposit the weld is quite obviously a big factor in the achievement of quality. For many applications, especially with low-strength structural steels, the welding operation is very tolerant of variations in the way in which a joint is welded. By contrast, when welding creep-resistant steels the properties of the completed joint can be critically dependent on the preheat level, heat input, run sequence, and post-weld heat treatment. In applications such as this, we need to know before starting to weld if the procedure we propose to use will give the required quality. We will certainly want this information for quality-control purposes, but frequently it is also required by some inspecting authority.

Broadly there are three groups of people who could be interested in the approval of a weld procedure. Firstly, the customer may have a member of staff or may appoint a representative who will be responsible for ensuring that the welding conforms with the requirements of the contract. In this context, the customer can be a private company or a government or nationalised organisation. Secondly, an official body may be involved in cases where the public safety is of concern when the fabrication is in service. A good example in the UK is the Civil Aviation Authority (CAA), which is responsible for checking the air-worthiness of civil aircraft. Finally, an insurance company will want the procedure to be approved by one of its surveyors before agreeing to insure the fabrication. In the UK, well known names such as Lloyd's Register of Shipping, Eagle Star, and National Vulcan are in this group.

When an approval is required, the weld procedure is first specified in terms of

a) welding process,
b) parent metal,
c) consumables – specification, care, etc.,
d) shop or site weld,
e) preparation (cleaning, edge preparation, assembly, etc.),
f) position of welding,

g) number and arrangement of runs,
h) electrical conditions,
i) gas shielding,
j) heat input,
k) preheat and interpass temperature,
l) back-gouging,
m) post-weld heat treatment,

This list is not exhaustive but indicates the detail with which a procedure should be specified.

A sample weld is now prepared using the procedure and is tested according to an agreed code. Any of the interested parties discussed above may have its own standards of acceptance or may agree with the manufacturer to accept a British Standard or other published national standard.

In general terms, the standard will call for the test plate to be examined by NDT, probably stipulating that a radiograph should be taken for record

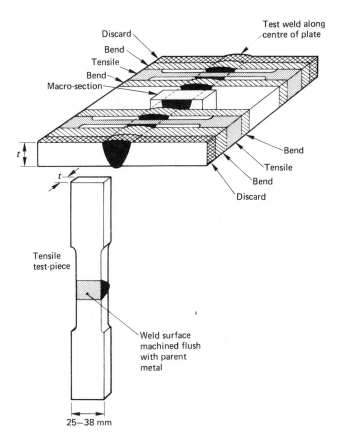

**Fig. 8.15** Method of cutting test-pieces from procedure-approval plate

purposes, and for test-pieces to be cut from the plate for destructive testing (fig. 8.15). Usually a transverse tensile test is made to demonstrate that the failure is in the parent plate, thus confirming that the correct electrode material has been selected. With aluminium and its alloys, the joint efficiency would be determined.

Bend tests are also a normal requirement, but their exact form will depend on the standard being used. The aim of the bend test is to place one part of the welded joint into tension with the object of confirming that the metal has sufficient ductility to bend round a specified radius and assessing if any defects present will initiate cracks. Three forms of test are currently used (fig. 8.16):

a) *face bend*, in which the top surface of the weld is in tension;
b) *root or reverse bend*, in which either the root of the weld or the reverse face of the weld is in tension;
c) *side bend*, in which a vertical slice is cut transverse to the weld so that the cross-section of the joint is in tension.

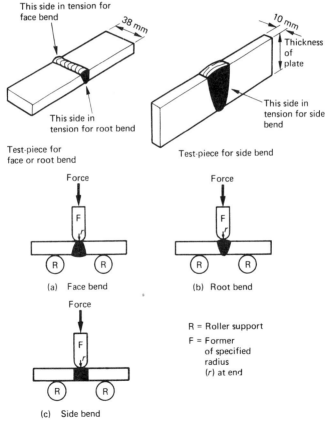

Fig. 8.16 Three types of bend test

In each case the test strip, which is prepared to the dimensions and finish specified in the standard, is placed across two rollers and a former is pressed against the opposite side to that which is being tested. The test-piece is forced downwards between the rollers, so that the lower surface is in tension, until the sides are parallel. The surface is then examined for cracks and tears.

Once the procedure has been approved, we must check that the welders can in fact produce welds to the quality represented by the test plate which was accepted by the inspecting authority. For this purpose the welder prepares another test plate similar to that used for the procedure approval. This is tested, perhaps not quite as exhaustively, and if it is satisfactory the welder is approved for work on the fabrication.

In some cases, rather than approving the welder for each job, it may be agreed that a general qualification will cover a range of work, provided the standard required is not higher than that achieved by the welder in the test. These two approaches can be seen in a set of documents prepared by British Standards:

BS 4870, 'Approval testing of welding procedures';
BS 4871, 'Approval testing of welders working to approved welding
        procedures';
BS 4872, 'Approval testing of welders when welding procedure approval
        is not required'.

Welder approval, which is often referred to as 'coding' of welders, can only tell us that under the conditions of the test the welder achieved a particular standard. It demonstrates the level of competence we can expect from the welder, but it would be wrong to assume that the same quality will always be achieved in welds made under production conditions – it simply provides a basis on which quality control of the actual welding operation can be established.

# 9 Shrinkage and distortion in fusion welding

## 9.1 Thermal expansion and contraction

Thermal expansion and contraction of metals are such familar phenomena in engineering that we often forget that without them many tasks could not be performed as readily, nor would many mechanisms function properly. Shrink-fitting of tyres on to wheels and sleeves on to shafts is practised regularly under closely controlled conditions. Thermal cut-outs rely on metal strips expanding in a predictable fashion. Conversely, the same phenomena can cause problems, as when overheated parts expand and foul their bearings. The physical laws governing the expansion and contraction of metals are well known and ample data are available to enable many applications to be analysed mathematically.

Since fusion welding involves a heating-and-cooling cycle, it is inevitable that thermally induced dimensional changes occur which have a profound effect on the behaviour of the joint during the joining operation. Unfortunately, the sequence of thermal events involved in welding is far from simple and is not readily amenable to mathematical analysis. On the other hand, it is possible to describe the contraction of a welded joint in a qualitative manner and to ascribe to the different stages empirical data established by observations made over a period of many years.

In practice we are principally concerned with the shrinkage that takes place during cooling after the heat source has passed along the joint line, although, of course, any restrained expansion resulting from heating before fusion may cause undesirable buckling, for example in thin sheets.

Shrinkage during welding is made up of three components:

a) contraction of liquid metal,
b) change of volume on solidification,
c) contraction of solid metal.

## 9.2 Solidification shrinkage

In practice it is difficult to separate the contraction of liquid metal and the change of volume on solidification, since melting and solidification in welding are progressing at the same time. Looking at a transverse vertical section through a welded joint during the solidification cycle (fig. 9.1), the two stages can be seen to interact. As the solidification front progresses upwards towards the centre-line, the solid metal occupies a smaller space than the liquid metal it replaces, i.e. its density is higher. The molten metal is also contracting, with the result that the surface of the pool should recede below the original level. However, at the same time, further molten

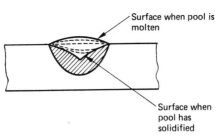

Fig. 9.1   Shrinkage during solidification

metal is being fed into the area by melting of the parent metal at the leading edge of the weld pool and by the addition of the melted electrode material. Thus, although appreciable shrinkage is occuring during cooling and solidification from the liquid state, the surface of the weld will not normally show evidence of this unless insufficient added metal has been available to compensate for the reduction in volume.

The end or termination of a weld run gives an interesting example of the importance of weld-metal feeding. If the heat source is suddenly removed, solidification proceeds as normal, but, since no further melting is taking place, the level of the surface falls and a crater is produced. In practice, the welder adds sufficient electrode or filler metal to build up the surface, possibly at the same time gradually reducing the heat input. This technique, known as crater filling, is of considerable importance in welding, since cracking can often be associated with an incorrectly or inadequately filled crater.

### 9.3  Contraction of solid metal

After welding, the weld metal is still hot and heat continues to flow into the parent metal until the whole joint has cooled to room temperature. During this period of cooling, the weld metal contracts both along the length of the joint and at right angles to it. This means that its volume should become smaller, but in fact this is prevented from happening by the restraining influence of the parent metal to which it is bonded. It is not easy to visualise this effect with an actual weld, because a number of events are happening at the same time, but we can illustrate the final results by considering the weld as a length of hot bar between two pieces of cold plate. As a starting point we can assume that, when it is hot, the bar fills the gap between the two plates. As the bar starts to cool, it contracts both in length and in cross-section.

### Longitudinal shrinkage

Some idea of the decrease in length can be obtained by using the equation

$$l_1 = l_0 (1 - \alpha \Delta\theta) \text{ or } l_1 = l_0 - l_0 \alpha \Delta\theta$$

where   $l_0$ = original length
$l_1$ = length after cooling through $\Delta\theta°C$

133

$\alpha$ = coefficient of linear expansion

$\Delta\theta$ = temperature change (°C)

The linear shrinkage or contraction is therefore equal to $l_0 \, \alpha \, \Delta\theta$.
For low-carbon steel, the following data would apply:

$\alpha = 14.3 \times 10^{-6}/°C$

$\Delta\theta$ = melting temperature − room temperature

$\quad = 1500°C - 20°C$

$\quad = 1480°C$

Suppose the weld is 1 metre long:

$l_0 \, \alpha \, \Delta\theta = 1000 \text{ mm} \times 14.3 \times 10^{-6}/°C \times 1480°C$

$\quad\quad\quad = 21.2 \text{ mm}$

This figure is only an approximation, since the value of $\alpha$ does not remain constant over the temperature range we quoted, but it does indicate that the contraction along the length of the weld should be appreciable. In practice the measured contraction is significantly less, being of the order of 1mm per metre length of weld. We must therefore examine why this should be so.

The first and most obvious factor is that the hot weld element is attached, i.e. bonded, to the cold plates on either side and is not free to shrink. The plates can be considered as being more or less rigid, so that, when the weld metal tries to contract, it is plastically deformed by being held to the original length. This means that tensile forces are being set up in the weld region, which are reacted by compressive forces in the plates (fig. 9.2).

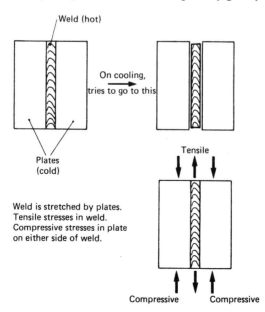

Fig. 9.2    Deformation of a weld-metal element during cooling

134

If these plates were perfectly rigid, when the cooling cycle had finished the welded joint would be the same length as the plates were originally. The compressive stresses induced, however, are of considerable magnitude and are in excess of the compressive yield stress of the parent metal. The result is that the plates themselves are plastically deformed, so reducing the overall length of the joint and thus accounting for the 1mm/m shrinkage quoted above. This is appreciably less than the amount of deformation which occurs in the weld element, principally because the yield stress of the relatively cold plate is higher than that of the hot weld metal. Some idea of the magnitude of the forces generated during shrinkage in welded joints may be obtained from the fact that, in low-carbon steel at the temperature involved, a compressive force of about 150 to 170 $N/mm^2$ is required to achieve a compressive strain of about $1 \times 10^{-3}$ mm/mm (i.e. a shortening of 1 mm per metre length of joint).

### Transverse shrinkage

Similar considerations apply when we look at shrinkage transverse to the weld, where the contracting weld metal tries to pull the plates towards the centre-line of the joint and as a result the whole joint area is in transverse tension. Again we have a situation where, because the hot weld metal has a lower yield stress than the cold plates, deformation first takes place in the weld but, at a later stage of cooling, as the relative yield stresses become more equal, some yielding of the parent material occurs and the overall width of the welded plates is reduced.

Strictly, the amount of transverse shrinkage which takes place depends on the total volume of weld metal, but as a general rule, for a given plate thickness, the overall reduction in width transverse to the joint at any point is related directly to the cross-sectional area of the weld. Similarly, as we would expect, the total shrinkage increases with the thickness of the plate, since the weld area is greater. It is possible to state the relationship in a general way:

$$\text{transverse shrinkage} = k\frac{A}{t}$$

where   $k$ = an empirical factor with a value between 0.1 and 0.17
         $A$ = cross-sectional area of weld
         $t$ = thickness of plate

This formula can be used to predict the shrinkage that will occur in a butt joint (fig. 9.3) and has been found to give good correlation with practical observations. In the case of a single-V butt joint the calculation can be simplified, since the ratio $A/t$ is equal to the average width (fig. 9.4) and the formula is reduced to

$$\text{transverse shrinkage} = k \times \text{average width of weld}$$

It should be noted that for a double-V weld the average width is not zero but is the value for *one* of the V's.

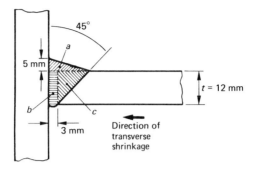

**Fig. 9.3** Estimation of transverse shrinkage in a 'T' butt joint

Transverse shrinkage = $0.1 \times \dfrac{A}{t}$

$A = a + b + c$

$\quad = \frac{1}{2} \times 5 \times (12 + 3) + (3 \times 12) + (\frac{1}{2} \times 12 \times 12)$

$\quad = 145.5 \text{ mm}^2$

$\therefore$ transverse shrinkage $= 0.1 \times \dfrac{145.5}{12}$

$\qquad\qquad\qquad\qquad = 1.21 \text{ mm}$

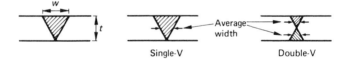

**Fig. 9.4** Transverse shrinkage in 'V' butt welds

Area of weld, $A = \frac{1}{2} \times w \times t$

Transverse shrinkage $= 0.1 \times \dfrac{A}{t}$

$\qquad\qquad\qquad\qquad = 0.1 \times \dfrac{\frac{1}{2} \times w \times t}{t}$

$\qquad\qquad\qquad\qquad = 0.1 \times w/2$

$\qquad\qquad\qquad\qquad = 0.1 \times \text{average width}$

*Angular distortion and longitudinal bowing*

Taking both longitudinal and transverse shrinkage, based on what has been said above the final shape of two plates welded together with a butt joint should be as shown in fig. 9.5(a). In practice, however, such a simple treatment does not apply, principally because the shrinkage is not distributed uniformly about the neutral axis of the plate and the weld cools progressively, not all at one time.

136

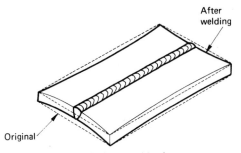

(a) Changes in shape resulting from shrinkage which is uniform throughout the thickness

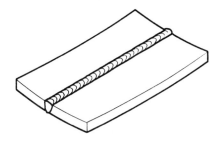

(b) Asymmetrical shrinkage tends to produce distortion.

**Fig. 9.5** Change in shape and dimensions in butt-welded plate

If we look at a butt weld made with a 60° included-angle preparation, it is immediately apparent that the weld width at the top of the joint is appreciably greater than at the root. Since the shrinkage is proportional to the length of metal cooling, there is a greater contraction at the top of the weld. If the plates are free to move, as they mostly are in fabricating operations, they will rotate with respect to each other. This movement is known as angular distortion (fig 9.5(b)) and poses problems for the fabricator since the plates and joint must be flattened if the finished product is to be acceptable. Attempts must be made, therefore, to reduce the amount of angular distortion to a minimum. Clamps can be used to restrain the movement of the plates or sheets making up the joint, but this is frequently not possible and attention has to be turned to devising a suitable weld procedure which aims to balance the amount of shrinkage about the neutral axis. In general, two approaches can be used: weld both sides of the joint or use an edge preparation which gives a more uniform width of weld through the thickness of the plate (fig. 9.6).

In the direction of welding, asymmetrical shrinkage shows up as longitudinal bowing (fig. 9.7). This is a cumulative effect which builds up as the heating-and-cooling cycle progresses along the joint, and some control can be

137

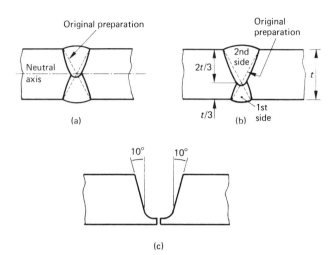

Fig. 9.6   Edge preparation designed to reduce angular distortion

  a) Double-V joints balance the shrinkage so that more or less equal amounts
     of contraction occur on each side of the neutral axis. This gives less
     angular distortion than a single 'V'.
  b) It is difficult to get a completely flat joint with a symmetrical double
     'V' as the first weld run always produces more angular rotation than
     subsequent runs; hence an asymmetrical preparation is used so that the
     larger amount of weld metal on the second side pulls back the distor-
     tion which occurred when the first side was welded.
  c) Alternatively, a single-U preparation with nearly parallel sides can be
     used. This gives an approach to a uniform weld width through the section.

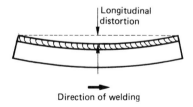

Fig. 9.7   Longitudinal bowing or distortion in a butt joint

achieved by welding short lengths on a planned or random-distribution basis
(fig. 9.8). Welding both sides of the joint corrects some of the bowing, but
can occasionally be accompanied by local buckling.

Angular distortion and longitudinal bowing are also observed in joints
made with fillet welds (figs 9.9 and 9.10). Angular distortion is readily seen

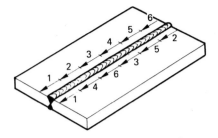

**Fig. 9.8** Sequences for welding short lengths of a joint to reduce longitudinal bowing

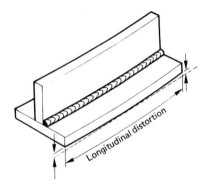

**Fig. 9.9** Longitudinal bowing in a fillet-welded 'T' joint

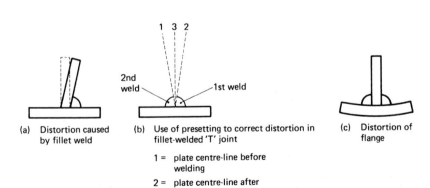

(a) Distortion caused by fillet weld

(b) Use of presetting to correct distortion in fillet-welded 'T' joint

2nd weld — 1st weld

1 3 2

1 = plate centre-line before welding
2 = plate centre-line after first weld
3 = plate centre-line after second weld

(c) Distortion of flange

**Fig. 9.10** Distortion in fillet welding of 'T' joints

139

in this case as a reduction of the angle between the plates and is greatest
for the first weld. Although the second weld, placed on the other side of the
joint, tends to pull the web plate back into line, the amount of angular
rotation will be smaller. With experience, the joint can be set up with the web
plate arranged so that the first angle is greater than 90° and thus ends up with
the web and flange at right angles. Even so, warping in the flange plate cannot
be ignored.

*Effect of heat distribution*
Finally, in our consideration of shrinkage and distortion we must not ignore
the importance of heat input. As we have seen in chapters 2 and 3, the heat
from the weld pool during solidification flows into the plate adjacent to the
fusion boundary. The width of metal heated to above room temperature is
greater than that of the fused zone, and the picture used above of a hot
weld-metal element between cold plates is an over-simplification. The heat
flowing into the plates establishes a temperature gradient which falls from
the melting point at the fusion boundary to ambient temperature at some
point remote from the weld.

The heated-band width is directly proportional to the heat input in joules
per mm length of weld and is therefore dependent on the process being
used. It follows that the amount of distortion and shrinkage will also vary
from one welding process to another. If the heat source moves slowly along
the joint, the heat spreads into the plate and the width of hot metal which
must contract is greater. The effect is less noticeable in thick plate but in
sheet material, say 2 mm thick, the differences are most marked. The MAGS
system, with its fast speed of travel, gives a narrow heat band compared with
the spread in oxy–acetylene welding, and it is possible to arrange the manual
processes in ascending level of distortion, i.e. MAGS, MMA, TAGS, and
oxy–acetylene welding.

9.4. Residual stresses
Solving the problems of distortion control during welding and determining
shrinkage allowances for design purposes are of such importance in fabrication that it is easy to overlook the fact that they are the products of plastic
deformation resulting from stresses induced by contraction in the joint. As
long as these stresses are above the yield point of the metal at the prevailing
temperature, they continue to produce permanent deformation, but in so
doing they are relieved and fall to yield-stress level. They then cease to cause
further distortion. But, if at this point we could release the weld from the
plate by cutting along the joint line, it would shrink further because, even
when distortion has stopped, the weld still contains an elastic strain equivalent to the yield stress. We can visualise the completed joint as an element
of weld metal being stretched elastically between two plates.

The stresses left in the joint after welding are referred to as residual
stresses. From our discussion of shrinkage and distortion, it can be seen that
there will be both longitudinal and transverse tension. In the case of the
longitudinal stresses, the weld itself and some of the plate which has been

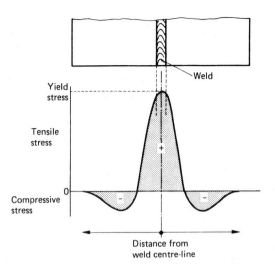

**Fig. 9.11** Distribution of residual stresses in a butt-welded joint

heated are at or near yield-stress level (fig. 9.11). Moving out into the plate from the heat-affected zone, the stresses first fall to zero. Beyond this there is a region of compressive stress.

It must be emphasised that all fusion welds which have not been subjected to post-weld treatments – in other words, the vast majority of welded joints – contain residual stresses. Procedures developed to minimise distortion may well alter the distribution of the residual stresses but do not eliminate them or even reduce their peak level. Having said this, since we cannot avoid the formation of residual stresses, it is appropriate to ask if we are worried by their presence. As with so many engineering situations the answer is not a simple yes or no. There are numerous applications where the existence of residual stresses would have little or no influence on the service behaviour of the joint – storage tanks, building frames, low-pressure pipework, and domestic equipment all provide examples of situations where the joints can be used in the as-welded condition without detriment.

If the service requirements do indicate that the residual stresses are undesirable, the designer must take them into account when selecting materials and deciding upon a safe working stress. This approach can be seen in the design of ships, where the combination of low temperatures and residual stress could lead to a type of failure known as brittle fracture. The designer selects a material which is not susceptible to this mode of failure even at the low temperatures which may be experienced during the working life of the ship; the presence of residual stresses is then not important. Similarly, in many structures subjected to loads which fluctuate during service – for example, bridges, earth-moving equipment, and cranes – the designer recognises the existence of residual stresses by choosing a working-stress range

141

which takes account of the role these stresses play in the formation and propagation of fatigue cracks.

There are, however, some specific applications where it is essential to reduce the level of residual stresses in the welded joint. With pressure vessels, because of the risk of a catastrophic failure by brittle fracture, stress-relieving is often a statutory or insurance requirement. Again, some metals in certain environments corrode rapidly in the presence of tensile stress, i.e. stress corrosion will occur. In these cases, a joint in the as-welded condition containing residual stresses suffers excessive attack; this is retarded if the joint is stress-relieved. Finally, when machining welded components, removing layers of metal near the joint may disturb the balance between the tensile and compressive residual stresses and further deformation or warping can occur. This can make it difficult to hold critical machining tolerances and it may be desirable in these circumstances to stress-relieve to achieve dimensional stability.

*Stress-relieving*

Various methods are available to reduce the level of residual stresses in welded joints. Heat treatment, overloading, and vibratory treatment can all be used, but the most common method is based on a controlled heating-and-cooling cycle, i.e. thermal stress relief. This technique makes use of the fact that the yield stress of a metal decreases as the temperature is raised. If a welded joint is heated to, say, $600^{\circ}C$, the residual tensile stress, which was equivalent to the yield stress at room temperature, is in excess of the yield stress of the metal at $600^{\circ}C$. Localised plastic deformation occurs, and the tensile stresses are reduced. At the same time, the compressive stresses which were in equilibrium with the tensile stresses are also reduced, to restore the equilibrium.

In stress-relieving practice, the temperature is raised until the yield stress has fallen to a low value at which residual stresses can no longer be supported. This clearly depends on the metal being treated, since the relationship between yield stress and temperature is critically influenced by alloy content, and this is reflected in the temperatures recommended in BS 5500:1985 for the stress-relieving of fusion-welded pressure vessels (Table 9.1).

Table 9.1 Stress-relieving temperatures for fusion-welded pressure vessels

| Type of steel | Stress-relieving temperature ($^{\circ}C$) |
| --- | --- |
| Low-carbon | 580−620 |
| Carbon−manganese | 580−620 |
| Carbon−½% molybdenum | 630−670 |
| 1% chromium−½% molybdenum | 630−670 |
| 2¼% chromium−1% molybdenum | 680−720 |
| 5% chromium−½% molybdenum | 710−750 |
| 3½% nickel | 580−620 |

If thermal treatment is to give a satisfactory reduction of residual-stress levels, it is important that differential expansion and contraction must not occur, otherwise new residual stresses will be induced. The heating and cooling must be carefully controlled so that the temperature is uniform throughout the component, and special furnaces equipped with comprehensive temperature-control systems have been designed for this purpose. In these furnaces the whole of the component or fabrication is heated, thus easing the problem of avoiding temperature gradients. Localised heating for stress relief is usually not recommended, especially with joints in flat plates, since there is always the risk of creating further stresses. In this connection, pipe welding poses particular problems. Stress relieving might often be desirable to reduce corrosion problems, but it would be impracticable to heat-treat a complete pipework installation. Local stress relief of pipe joints in situ is, therefore, allowed by some authorities, provided that the temperature distribution is controlled. This is usually achieved by specifying the minimum temperature at the joint line and at some specific point remote from the weld – a typical example is shown in fig. 9.12.

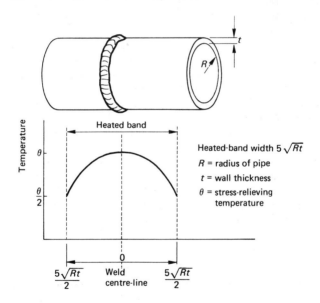

**Fig. 9.12** Typical specification for temperature distribution during local stress relief of welded butt joints in pipe

# 10    Mechanising arc welding

## 10.1 Factors affecting welding speed

Welding is used in so many widely differing industries that it is difficult to assess the relative usage of the various processes. However, it is probably not too inaccurate to suggest that the manual processes discussed in chapters 5 and 6 account for about 90% of all the fusion welding of plates, pipes, sections, castings, and forgings carried out in the UK. To some extent this reflects the fact that much of the work undertaken benefits from the versatility of the manual systems. Relatively short runs, constant changes in the position of welding, movement across the shop floor from one fabrication to another, and a mix of different types of work or joint all encourage the use of manual welding, since the welder's skill can be deployed to the best advantage in coping with the varying requirements. On the other hand, some fabrications, especially where long joints are involved, could be designed to take advantage of higher speeds of welding or greater deposition rates than those obtainable with manual techniques.

High welding speeds involve more than just travelling faster along the joint. If a sound bond is to be achieved in the weld, the minimum heat input for fusion must always be exceeded. Therefore, the rate of heat input to the arc must be increased as the travel speed is raised. This follows from the relationship given in chapter 3:

$$\text{heat input to weld pool (kJ/mm)} = \frac{\text{current (A)} \times \text{arc voltage (V)} \times 60}{\text{weld travel speed (mm/min)} \times 1000}$$

To achieve high welding speed we must use a high current, as we cannot make large changes in arc voltage. Similarly, to deposit significantly larger weld runs than those produced in the normal operating range for manual welding (100 to 350 A) the current must be increased to between 500 and 1000 A to give more rapid electrode melting (fig. 10.1). Currents of this magnitude are not easily managed manually, and we must look at some form of mechanisation which takes from the welder the tasks of controlling the arc length, feeding the electrode, and traversing the joint.

## 10.2 Mechanised TAGS welding

Probably the easiest system to mechanise is TAGS welding, especially for use on thin sheet where filler metal is not required. In essence, all that is needed is a method of moving the TAGS torch along the joint line at a fixed height above the work, thus keeping the arc length constant.

This assumes that the work is flat, at least along the joint line, to within the limits allowed for variations in arc length. More often than not when

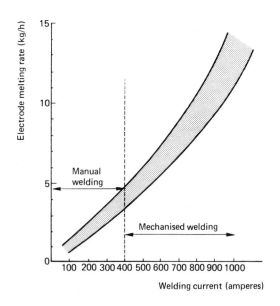

**Fig. 10.1** Relationship between welding current and electrode melting rate in arc welding

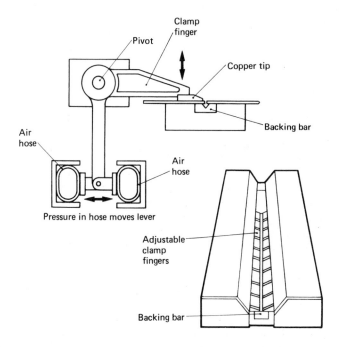

**Fig. 10.2** Longitudinal clamping system for mechanised TAGS welding

145

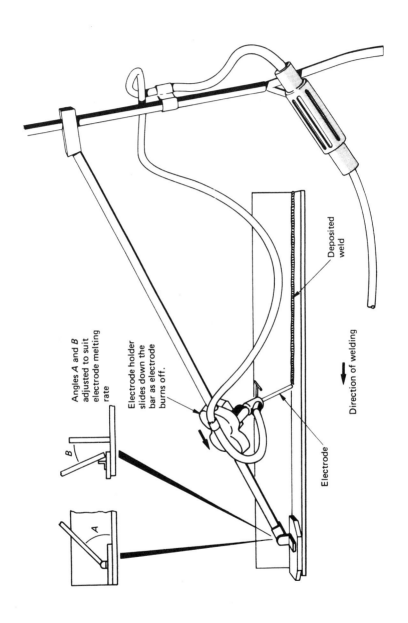

Angles A and B adjusted to suit electrode melting rate

Electrode holder slides down the bar as electrode burns off.

Deposited weld

Direction of welding

Electrode

B

A

**Fig. 10.3** Gravity welding

146

sheet is being welded by the TAGS process, the abutting edges lift and distort under the influence of the heat from the arc. In manual welding, the operator can allow for this and make adjustments to keep the arc length constant. For mechanised welding, the sheets must be clamped to prevent the edges lifting, and special jigs of the type shown in fig. 10.2 need to be used. These are fitted with a backing bar against which the sheets are clamped. The bar is usually made of copper, since this helps to conduct the heat away from the weld area. The clamping force is provided by a series of 'fingers', each about 75 to 100 mm wide and fitted with copper tips which, by conducting heat away from the weld, help to keep the heated width as small as possible, thus minimising distortion.

Where arc-length control is critical, it is possible to hold the torch in a clamp which can be raised or lowered in response to commands from a servo-control system linked to the arc voltage.

## 10.3 Gravity welding

In general, it is not easy to mechanise manual metal arc welding. The short length of the electrodes and the difficulties of controlling the arc length have hampered the development and exploitation of suitable systems. One technique which has been successfully used in a number of applications is gravity welding (fig. 10.3). The electrode is clamped in a holder which can slide up and down a bar or rod held at a predetermined angle to the joint. The joint is almost invariably a 'T', and the top of the electrode is positioned in the root of the joint. Once the arc has been initiated, the end of the electrode melts, the electrode gets shorter, and the holder moves down the bar, keeping the tip of the electrode in contact with the joint. The weld is thus deposited until the electrode has been reduced to about 50 mm, at which point the movement of the holder ceases and the arc is extinguished. A fresh electrode is placed in the holder, the slider bar is moved along the joint, and the weld is restarted at the point at which the previous electrode stopped.

The successful operation of gravity welding depends on two factors. Firstly, the electrode must have a flux which melts in the form of a cone at the tip (fig. 10.4). By resting the outer edge of the flux cone on the plate in the joint, the arc length is controlled by the melting of the flux. Secondly,

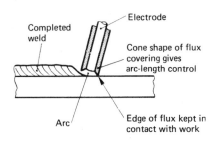

Fig. 10.4  Arc-length control in gravity welding

147

the angle between the slider bar and the joint must match the melting rate of the electrode if the system is to be stable.

The improvement in deposition rate over manual welding is small, but, by using one operator to look after, say, four or five gravity welders, the overall output can be appreciably higher. The system is best suited to depositing fillet welds in the horizontal position, and a typical application is welding stiffeners to plates. In spite of its simplicity, gravity welding is still not widely used.

## 10.4 Continuous-electrode systems

Undoubtedly the most effective way to mechanise arc welding is to use a continuous electrode coupled with a suitable method of arc-length control.

We have already seen a continuous electrode used with a self-adjusting arc in MAGS welding (section 4.6), and many mechanised systems are based on this principle, but it will be remembered that self-adjustment works well only when fast electrode feed rates are used, such as would be the case with small-diameter electrodes. There are considerable advantages to be gained, however, by using larger diameter wires; for example, higher currents and deposition rates can be achieved at slower wire feed speeds, thus simplifying the construction of the wire-drive unit. In some ways this situation is then similar to that encountered in MMA, where the welder controls the arc length by adjusting the rate at which the electrode is moved towards the weld pool. With mechanised welding, the electrode is fed into the arc area by means of a motor and a drive-roll system. We can, therefore, adopt the MMA system and vary the electrode speed to keep the arc length constant.

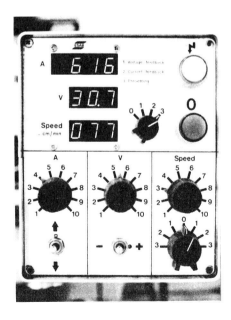

**Fig. 10.5** Control panel for a submerged-arc welder

148

The voltage, current, and travel speed are set on the control panel (fig. 10.5). Current is supplied by a drooping-characteristic power-supply unit (fig. 10.6) which keeps the current at the predetermined value. The motor feeds the electrode at a speed which gives the desired arc length at the current being supplied by the power unit. The voltage corresponding to this arc length is used as a reference against which the actual arc voltage measured during the welding operation is compared. If the arc length increases, the voltage rises. The comparator sends a command signal for the motor control to increase the speed in proportion to the difference between the arc and the reference voltages. Since the burn-off rate of the electrode is determined mainly by the current and varies only slightly with voltage, the electrode is now moving faster than it is melting and the tip approaches the weld pool, thus reducing the arc length to its original value. At the same time, the arc voltage falls back to the reference voltage and, as the arc approaches its original length, the electrode feed rate slows down until it also returns to the preset value.

Table 10.1 summarises the two systems used in mechanised arc welding.

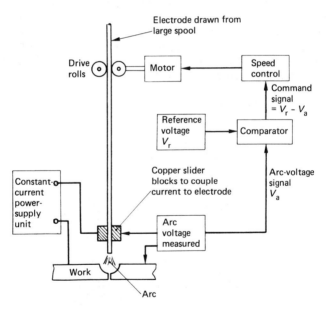

**Fig. 10.6** Typical layout of a controlled-arc system for consumable-electrode welding

### Consumables in mechanised arc welding

Two types of electrode are commonly in use with mechanised arc welding: solid bare wires and flux-cored electrodes.

**Table 10.1** Mechanised arc-welding systems

| System | Electrode dia. (mm) | Motor drive | Power supply |
|---|---|---|---|
| Self-adjusting arc | 0.8 to 3.0 | Constant speed at preset value to determine current | Flat characteristics with voltage selector |
| Controlled arc | 3.0 to 6.0 | Variable speed; servo control based on arc voltage | Drooping characteristics with current selector |

**Solid bare wires**   These are of a composition similar to that of the parent plate, but must contain sufficient alloying elements and deoxidants to allow for losses in the arc. For example, to achieve 1.0% manganese in a steel weld-metal deposit, the wire may need to contain 1.8% of this element.

Steel wires are usually copper-coated to give better electrical contact with the copper current slides and are used in diameters from 0.8 to 6 mm.

**Flux-cored electrodes**   By placing the flux in the centre of a tubular electrode, these provide an answer to the problem of current pick-up experienced with continuous-covered electrodes. The two most commonly used diameters are 2.4 and 3.2 mm. The electrode may be a simple tube or it may have a more complex cross-section, such as one of those shown in fig. 10.7.

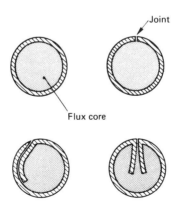

**Fig. 10.7** Cross-sections of typical flux-cored electrodes

Fluxes have not yet been classified in the same way as those used in MMA electrodes, but a typical flux could contain some calcium fluoride and carbonate together with a deoxidant, the exact composition depending on the manufacturer. The flux usually constitutes about 20% of the total mass of the electrode.

Flux-cored electrodes are particularly useful in the deposition of hard-wearing surface layers. Usually these deposits contain large amounts of alloying elements such as manganese, tungsten, cobalt, and chromium, and it is often difficult to produce a suitable composition in wire form. It is convenient, therefore, to use a carbon steel for the tubular part of the electrode and to pack the core with the alloying elements. As the electrode melts, the alloys mix with the steel and a metal of suitable composition is deposited on to the plate being surfaced.

*Shielding systems for mechanised welding*
The electrodes described can be used with either a flux or gas shield. These may provide the only means of shielding the weld pool from atmospheric contamination or may be used simply to supplement flux already incorporated into the electrode.

Bare wires are invariably used with either a gas shield (MAGS welding) or a flux which is preplaced along the joint line (submerged-arc welding).

Mostly, flux-cored electrodes require the additional protection provided by a gas shield and as such can be seen as a variant of MAGS welding. Flux-cored electrodes have also been developed which can produce their own shield by decomposition of the flux in the core.

## 10.5 Practical welding systems
The various combinations of electrode type and shielding system give a number of possible mechanised processes, as shown in Table 10.2.

Submerged-arc welding is the mechanised process most commonly found in fabricating and welding shops. It is used to weld pressure-vessel shells, box- and I-section girders for bridges, stiffened panels for ships, longitudinal

**Table 10.2** Mechanised welding systems

| Electrode type | Shielding method | | |
| --- | --- | --- | --- |
| | Flux | Gas | None |
| Solid bare-wire electrode | Submerged-arc welding | MAGS welding | |
| Flux-covered electrode | Hard-facing systems | Flux-cored arc welding (MAGS) | Self-shielded flux-cored welding |

butt joints in the manufacture of pipes, and many other products where long straight joints are used (see page 35).

A number of alternative traversing mechanisms can be used. The main requirement is that the method selected should be capable of moving the complete welding head − i.e. wire feed, electrode spool, and control box − at a stable controllable speed along the joint line. Some typical examples from current practice are

a) hand-operated trolley, sometimes called a deck welder;
b) tractor unit, electrically driven along a portable track;
c) motorised drive along a horizontal beam;
d) screw-fed saddle on a lathe-type bed;
e) hydraulic ram.

Alternatively, the welding head can remain fixed and the work be moved beneath it. This technique is frequently used when welding large-diameter pipes or shells.

With submerged-arc welding, a strip electrode which gives wide deposits on flat surfaces can be used where a large surface is to be covered with a dissimilar metal (fig. 10.8). For example, the inside of a steel chemical vessel can be covered in this way with a stainless-steel weld deposit to improve corrosion-resistance.

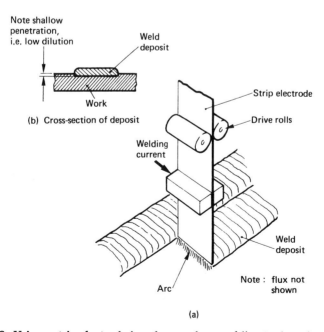

**Fig. 10.8** Using a strip electrode in submerged-arc welding to deposit surface layers on to plates

152

## 10.6 Advantages in using mechanised arc welding

From our discussion of the ways in which arc welding can be mechanised, five clear advantages for its use can be identified:

a) It offers the ability to use high welding currents, which means that faster welding speeds can be employed.

b) It is possible to weld longer lengths without stopping.

c) Because the traverse speed, arc voltage, arc current, and wire-feed speed are all controlled, the welds can be deposited with more consistency than in manual welding, i.e. there is better reproducibility. This can be a considerable attraction where quality is to be controlled by monitoring welding parameters.

It should be noted that the benefits of reproducibility can only be achieved if the input to the welding operation is consistent. In other words, the fit-up must be accurate: joint gaps uniform, bevel angles consistent, and plates aligned correctly. At the same time, the full implication of reproducibility must be recognised. If the correct welding conditions and fit-up are chosen, the welds can be reproducibly good; but if, say, the heat input is set wrongly at the start, lack of fusion may be reproduced along the length of the weld.

d) With high welding speeds it is possible to achieve smaller heated-band widths, accompanied by a reduction in the amount of shrinkage and distortion.

e) As the current is increased, not only does the deposition rate become larger but the weld also penetrates deeper into the plate. This can be a distinct advantage in the welding of 'T' joints (fig. 10.9).

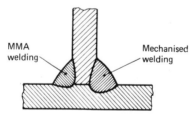

**Fig. 10.9** Comparison of penetration in welds of the same leg length deposited by MMA and mechanised welding

## 10.7 Limitations of mechanised welding

At high currents, the weld pool is large and we have difficulty in controlling the molten metal unless it is held in place by gravity. Admittedly fillet welds can be deposited in the horizontal–vertical position, but there is always a tendency for the metal to run on to the horizontal plate and form an overlap. We can restrict the current and/or weld size in horizontal–vertical welding to avoid this defect, but the degree of success depends on the type of flux, if any, which is used. It follows that, for the best results from mechanised arc welding, the work must be manipulated so that welds are deposited in the flat position.

The high currents used in mechanised welding raise problems in depositing root runs. The deep penetration achieved makes it difficult to control the melt-through, and, without some form of support, uniform-penetration beads are not usually a practical proposition except with mechanised TAGS welding. This means that, when butt welding a plate, it is preferable to weld from one side first, turn the plate over, gouge a groove along the joint line, and then complete the joint from the second side. Where it is essential that the welding is from one side only, various methods of backing may be used to support the weld, some of which are illustrated in fig. 10.10.

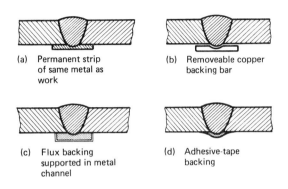

(a) Permanent strip of same metal as work

(b) Removeable copper backing bar

(c) Flux backing supported in metal channel

(d) Adhesive-tape backing

**Fig. 10.10**  Methods of backing joints during mechanised welding

## 10.8  Electro-slag welding

In all the mechanised processes considered so far, we have adopted the conventional practice of using an arc to establish a weld pool and then moving the heat source along the joint line to give progressive fusion and solidification. Where the thickness of the plate rules out the possibility of completing the weld in one pass, plate-edge preparation and multi-pass techniques are used. A completely novel approach to the welding of thicker plates is to place the joint vertical with a gap of, say, 25 to 50 mm between the square edges of the two plates and to fill the gap with molten metal (fig. 10.11). Water-cooled copper shoes must be provided on both sides of the plates, to prevent the molten metal from running out of the joint, and these shoes are moved up the joint as the gap is filled.

In many ways the system is similar to continuous casting, with the important exception that the 'cast' metal fuses to the plates making up two sides of the 'mould'. To achieve fusion of the plates, which is necessary for good bonding, the heat source must operate in the gap itself. An arc can be used for this purpose, and a flux-cored electrode with a gas shield

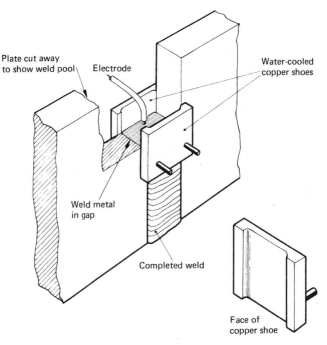

Plate cut away
to show weld pool

Electrode

Water-cooled
copper shoes

Weld metal
in gap

Completed weld

Face of
copper shoe

**Fig. 10.11** Schematic diagram of the arrangement for electro-slag welding

(electro-gas welding) has been used successfully on plates from 12 to 50 mm thick.

It is more common, however, to use the heat generated by the passage of a current through a molten slag bath (fig. 10.12). When solid, welding fluxes are non-conductive. On the other hand, when the flux or slag is molten, it conducts current but has a significant electrical resistance. Appreciable amounts of heat are generated when a high current flows in a molten slag bath positioned between the two plates, and both the parent metal and the electrode will be melted. Slag-bath heating is used in electro-slag welding and has been successfully applied to the welding of plates from 20 mm up to 350 mm thick.

The welding operation is begun on a piece of scrap plate attached to the bottom end of the joint (fig. 10.13). An arc is struck between the electrode and the starter plate. Granular flux is added and is melted by the arc. As soon as the gap is filled with molten flux, the arc is automatically extinguished and resistance heating begins. As the electrode melts, it fills the gap with metal which mixes with the fused plate material, and the surface of the weld pool rises up the joint. Once equilibrium conditions have been established, solidification at the bottom of the pool proceeds at a steady rate, and the weld pool is of constant depth. Some of the slag

155

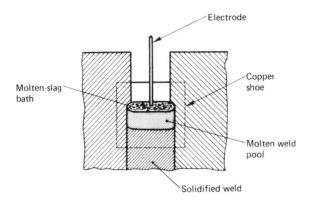

Fig. 10.12  Diagrammatic section through an electro-slag-welding system

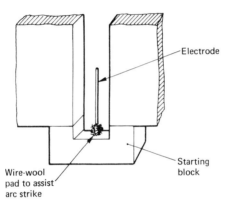

Fig. 10.13  Starting an electro-slag weld

coats the faces of the copper shoes and the surface of the solidified weld, and it is necessary to add small amounts of granulated flux from time to time to compensate for this loss.

Two electrode-feeding systems are in use at present. In electro-slag welding, one or more electrodes are introduced into the gap just above the surface of the slag bath, and the complete wire-feed unit is moved up the joint as the gap is progressively filled with weld metal (fig. 10.14). The alternative is to feed the electrode down a long tube positioned in the gap before welding begins. With this variant, known as consumable-guide welding, the wire-feed unit is positioned at the top of the joint and is not moved during the welding operation (fig. 10.15). The latter system is mechanically simpler and is suitable for welding shorter lengths of joint. The guide, which can take various forms, is melted into the joint.

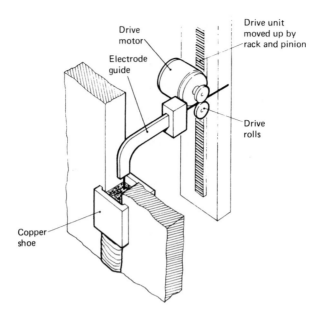

**Fig. 10.14** Electro-slag wire-feed unit

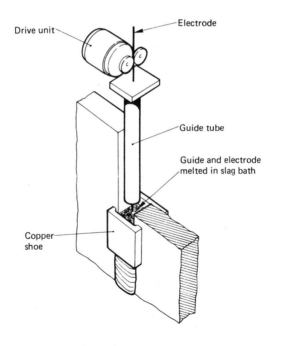

**Fig. 10.15** Wire feed for consumable-guide welding

The first 25 mm or so of the weld when the arc was being used to establish a slag bath is invariably defective, containing oxides and porosity. The starter plate is, therefore, designed so that it can be cut off, leaving sound metal at the start of the actual joint. It follows that, once an electro-slag or consumable-guide weld has been started, it should not be stopped until the top of the joint has been reached, as any attempt to restart the system will leave a volume of defective weld metal in the joint.

The weld produced by the electro-slag process tends to be slightly barrel-shaped in cross-section (fig. 10.16) and is relatively free from both transverse and longitudinal distortion. Shrinkage will still occur, of course, but, since distortion is negligible, this can be predicted and allowed for. This feature of electro-slag welding has made it particularly attractive in shipbuilding, where vertical joints in the hulls of large tankers have been successfully welded. The process is also used in the fabrication of thick-walled pressure vessels, but here there may be a need to post-weld heat-treat the structure to restore the notch ductility which may have been lost in the joint area due to the very slow cooling cycles in the heat-affected zone.

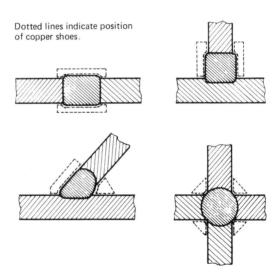

Dotted lines indicate position of copper shoes.

**Fig. 10.16**  Typical configuration for electro-slag-weld joints

Typical welding speeds are 1 to 3 m/h, depending on thickness. These may at first sight seem to be very low, but, when it is realised that they are for completed welds on plates in excess of 30 mm thick, electro-slag welding is seen as one of the fastest processes available to the fabricator.

## 10.9 Robots for welding

Mechanised welding is at its best on straight-line joints, whether butt or 'T'. On some equipments it is possible to steer the welding head so that it follows gentle curves. We can also move the head up and down to allow for changes in the vertical plane. Both of these operations involve the use of templates or optical guiding systems. Sensing devices may also be incorporated to check where the head is in relation to the joint line.

If there are a number of changes of direction along the joint, mechanised welding becomes difficult to set up and control. Frequently the answer is to stay with manual welding and provide the welder with a manipulator or other aids designed to increase the amount of time spent welding as opposed to setting up, tacking, moving around the component, and so on.

Robots offer an alternative to both manual and mechanised welding.

### What is a robot?

Robots take many forms but are essentially mechanised devices which can repeat a sequence of operations with a high degree of consistency. They may be used for a simple operation such as palletising or machine-loading, or they may be programmed to perform complex assembly tasks.

In a welding shop, a robot is a manipulating tool which moves the welding head along the joint line but at the same time maintains the correct relationship between the welding gun and the work. This means altering the gun or torch angle, keeping the nozzle-to-plate distance constant, and introducing a weaving movement when needed. The robot can do this only if it is given instructions, so there must be a facility to program the complete welding sequence, including control of the welding parameters. Ideally, there should be some form of feedback, so that the robot can check that it is welding in the right place.

### Types of robot

There are three principal types of robot used for welding operations. They are distinguished by the way the arm moves to position the welding torch or gun in relation to the joint line.

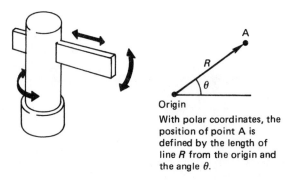

With polar coordinates, the position of point A is defined by the length of line $R$ from the origin and the angle $\theta$.

**Fig. 10.17** Polar-coordinate robot

159

**Polar-coordinate machines**  These are similar to the gun-and-turret mounting on an army tank (fig. 10.17). The vertical column ('turret') can rotate about its axis, and the arm can move in or out. At the same time, the end of the arm can be moved up or down by pivoting at the junction of the arm and the column, just like a gun in a tank.

**Cartesian-coordinate machines**  These position the gun by using three movements: left–right, up–down, and in–out (figs 10.18 and 10.19). In geometric terms these are the $x, y$, and $z$ directions. Sometimes these machines are called 'rectangular-coordinate machines'.

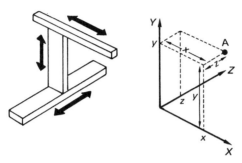

With Cartesian coordinates, the position of point A is defined by measurements $x$, $y$, and $z$.

Fig. 10.18  Cartesian-coordinate robot

Fig. 10.19  Typical coordinate welding robot

**Jointed-arm machines**   These are closest in operation to the human frame. They have a rotating turret (the human waist) and several joints in the arm which may replicate the shoulder and elbow (figs 10.20 and 10.21). This configuration makes a jointed-arm machine the most flexible of all the types of robot, but it requires the most complex control system.

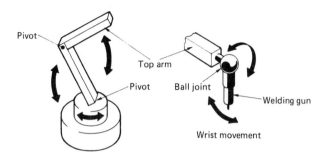

Fig. 10.20   Jointed-arm robot

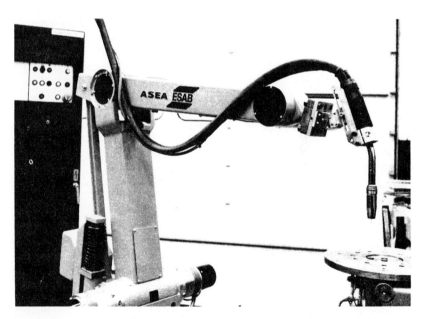

Fig. 10.21   Typical jointed-arm robot

All machines can have a wrist joint attached to the end of the arm, to allow the welding gun to be correctly positioned.

### Degrees of freedom

Looking at these different robots, we can readily see that there are a number of independent movements. Some of them give rectilinear motion, i.e. in a straight line. Others are rotational in character.

Each of the three machines described above has three basic axes. The addition of a wrist movement adds two more axes of rotation giving five in all, which is the minimum required for most welding operations. We can extend the capability of a robot either by moving it bodily along a track or by holding the work in a manipulator which may add one, two, or three axes of motion.

### Seam tracking

So far, robots have been mainly used with three welding processes: TAGS and MAGS (which we have already discussed in chapter 6) and resistance welding (which we looked at briefly on page 36 but will return to in the next chapter).

Resistance welding is particularly well suited to robotic applications, as here the robot's prime task is to position the gun with the electrodes in alignment. The spot weld is completed with the robot stationary.

With both TAGS and MAGS welding, the robot moves while welding. The pattern of movement can be complex. At first sight the branch connection in fig. 10.22 seems to be straightforward, but closer inspection shows that the welding gun not only moves along the joint line but also points in different directions. The axis of the gun for a number of positions is indicated by the centre lines of the nozzles.

**Fig. 10.22** Changes in orientation of the gun axis when welding a branch connection

This sort of movement is not too difficult with a TAGS torch. The principal requirements are to position the tip of the electrode along the line of the joint and to maintain the arc length within ±0.5mm.

However, considerable difficulty may be experienced if we need to make filler-metal additions with TAGS welding. The wire must be introduced into the leading edge of the weld pool and must be along the centre line of the weld. This introduces yet another movement which must be incorporated into the wrist action. Welding into corners poses particular problems, as it becomes impossible to accommodate the wire-feed tube ahead of the weld pool.

MAGS is the arc process most commonly used with robots. It is particularly easy to adapt MAGS welding to robotics, as the arc length is controlled from the power-supply unit and small variations in nozzle-to-plate distance can be tolerated. It is also possible to angle the torch when welding into corners, thus avoiding the problems associated with wire-feeding in TAGS welding. However, provision does have to be made in MAGS welding to ensure that the wire-feed hose can move freely, and attention needs to be paid to the problems caused by spatter blocking the nozzle and restricting the gas flow. These become particularly important when operating towards the outer limits of the working envelope of the robot. (The working envelope represents the furthest points which can be reached by the end of the arm.) When the arm is fully extended, it may not be possible to position the torch so that the correct welding angle or weaving motion can be maintained.

### Controlling a welding robot
The robot must be told what to do. It needs instructions on where to start and stop, what path to follow, and how the torch must be manipulated. To some extent separate from the operation of the robot but nevertheless an

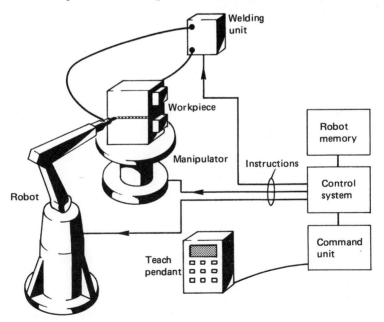

**Fig. 10.23** Control system for a welding robot

163

integral part of the operation, the welding parameters must be controlled and varied as necessary. These instructions are stored in a memory unit (fig. 10.23).

The memory can be programmed in a number of ways. In most current applications of welding robots, the memory is instructed via a command unit called the teach pendant. This typically has a collection of keys, rather like a large calculator, and it forms the interface between the human operator and the robot (fig. 10.24).

Fig. 10.24 Teach pendant for a jointed-arm robot

The simplest set of instructions identifies points where a weld is needed. Resistance spot welding is a good example of this type of operation. The robot is driven by the operator to the weld location. This position is recorded, and the robot is driven to the next location, which is also logged in the memory. The procedure is repeated until the positions of all the welds have been noted. The instructions are completed by adding the welding commands for each point. The program is stored in the memory as a series of numbers. When the robot is in operation, this data is supplied to the control unit, which responds by issuing instructions to the various moving parts and to the welding unit.

When a robot is being used for resistance welding, we are usually not too worried about the precise path it takes between spot welds. With arc welding, the situation is different (fig. 10.25). The start and stop positions for the weld run can be fixed in the same way as for resistance welding, but the trajectory or path between these two points must be precisely defined as it controls where the weld is deposited. The path can be a straight line or one of a number of predetermined curves.

164

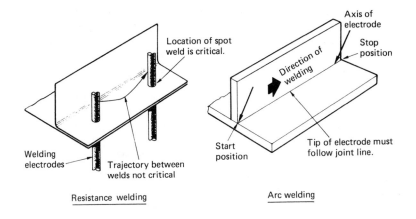

Fig. 10.25 Trajectories in resistance and arc welding

If the joint line is not easily defined in mathematical terms, e.g. a straight line or an arc of a circle, or has a number of changes of direction, the robot can be driven along the path it is to follow when welding. The coordinates are read at short intervals and again stored in the memory. If there are sudden or frequent changes of direction, allowance has to be made for the possibility of the robot overshooting.

As an alternative to using the teach pendant as a means of inputting data to the robot memory, a welder's movements can be automatically recorded. This is of particular value where the gun has to be manipulated, for example to produce a variable weaving motion. In a case like this, it is convenient to let an operator weld the joint using a gun fitted with a number of sensors which monitor all the critical movements. The data from these sensors is stored in the memory and later used to instruct the robot to produce a weld which copies the one made by the welder.

The above methods of giving instructions are called 'on-line' programming and reflect current practice with most arc-welding applications. Although they give good results, the robot cannot be used for production while it is being programmed, and this can be a lengthy operation. A more satisfactory method would be 'off-line' programming, where the instructions can be written away from the robot and stored on a tape or disk ready to be fed into the memory when needed for production. At present, off-line programming of welding robots is handicapped by the lack of a standard language for the instructions and the need to fine-tune the program on the robot to accommodate inaccuracies in set-up.

### Robots – yes or no?
The decision whether or not to install robots is markedly influenced by factors remote from the technological aspects which we have been discussing. A robot system represents a large capital investment, two to three

165

times the cost of the robot itself, which must be evaluated. It requires changes both in management structures and work-force practices. The effects are not confined just to the welding shop — practices in associated design, fabrication, and inspection operations need to be modified to respond to the new demands made by the robots. Staff must be retrained, and production will inevitably be interrupted.

Against all this, the use of welding robots offers significant advantages. Although there may not be increases in welding speed, output per station can be much higher — typically two to four times that of a manual station. Robots can be worked almost continuously throughout a shift, so there is less idle time than with manual operations. Also, unlike human beings, the robot can work 24 hours a day. A major attraction is the greater consistency of the product — this can be an important factor in quality assurance.

The successful introduction of robots depends on a number of technical factors. For example:

a) Joint accuracy must be better than that usually tolerated for manual welding. The welder can accommodate variations in root gap or face by adjusting such things as travel speed, arc length, electrode angle, and weave pattern. The robot is not able to respond so readily — if at all — and would probably give variable penetration or even burn through.
b) Component design must allow for access by the robot arm. Can the gun be brought to the joint with the correct angle and/or nozzle-to-plate distance?
c) If a jig is needed for assembly, it must not interfere with the movement of the robot arm.
d) Distortion control often involves extra movements which can complicate the programming of the robot.
e) Planned maintenance is absolutely essential and may well be novel in a traditional welding shop which uses only manual welding.

# 11    Resistance welding

In everyday life we use a whole range of equipments fabricated from sheet metal. Cars, washing machines, refrigerators, radiator panels, and storage cupboards are but a few of the items which we have come to accept as an integral part of modern living. The fact that they are available at reasonable cost to a large number of people is the direct result of developments in engineering techniques for the rapid production of identical units. A major factor has been the ability to make joints quickly and with minimum distortion.

A brief look at a mass-produced sheet-metal component highlights a dependence on lap joints. These are attractive as they can be secured in a variety of ways, some of which were mentioned in chapter 1. Pop rivets, self-tapping screws, soldering, and adhesive bonding are all used. The manufacturer thus has considerable freedom of choice and can concentrate on production-engineering aspects, but, where loads are to be transmitted, resistance welding in one of its different forms is often the first choice.

## 11.1  Basis of resistance welding
Resistance welding depends on the heating effect of a current flowing through an interface between two overlapping sheets (fig. 11.1). The interface offers a resistance to the flow of the current, and the energy expended is converted to heat. Applying Ohm's law, the voltage ($V$) required for a current flow ($I$) is given by $V = IR$, where $R$ is the resistance of the interface. The total energy for a current flow lasting $t$ seconds is expressed as

$$H = IVt = I(IR)t = I^2RT \text{ joules}$$

The heat so developed is concentrated into the localised area where the sheets are brought into intimate contact as a result of pressure applied by the electrodes. As the current continues to flow, the temperature rises until melting occurs at the interface, forming a weld pool. If the current is now stopped, the joint area cools and solidification takes place under pressure. The resultant weld nugget provides a localised bond between the sheets which can transmit a load through the joint.

The strength of the weld depends on the cross-sectional area of the nugget at the joint line. With small diameters under tensile tests, these welds shear along the interface (fig. 11.2). As the diameter is increased, the load at which such a failure occurs also gets larger. A point is reached where the mode of failure changes and fracture takes place in one of the sheets around the periphery of the weld, leaving the nugget or slug attached to the other sheet.

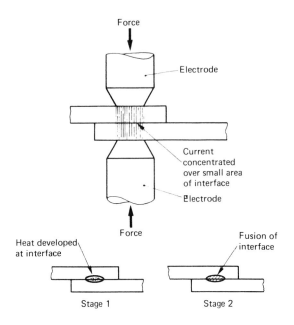

Fig. 11.1 Principles of resistance spot welding

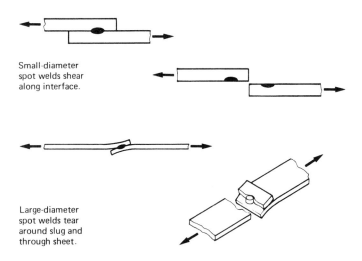

Small-diameter spot welds shear along interface.

Large-diameter spot welds tear around slug and through sheet.

Fig. 11.2 Failure of spot welds subjected to load

168

### Spot-welding sequence

A typical machine for resistance spot welding (fig. 2.40, page 37) contains a transformer which reduces the mains voltage to the working voltage (4 to 25 V), a means of controlling the current, and a timer, in addition to a mechanical system for applying force to the electrodes.

Welding sequences can vary, but the simplest (fig. 11.3) would be as follows.

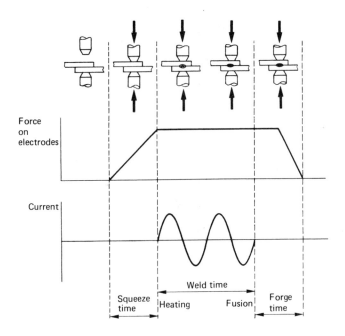

**Fig. 11.3** Typical spot-welding cycle

a) The joint is placed between the electrodes. Care must be taken to align the work with the electrodes, to ensure that the weld is in the correct position. Jigs are frequently necessary for this purpose if the high throughput rates of which resistance spot welding is capable are to be achieved.
b) The weld sequence is started, usually by operating a foot switch or a pedal.
c) The electrodes close, gripping the work and bringing the sheets into contact.
d) The pressure is increased to an optimum level over a period known as the squeeze time.
e) The current is switched on and is allowed to flow for a pre-set period during which melting occurs and the weld grows to the required size. In general, weld times are appreciably less than one second and are quoted in cycles — one cycle equals $\frac{1}{50}$ second, i.e. the inverse of the frequency of

169

the standard a.c. supply in the UK. As an example, for sheet steel (0.6 to 1 mm thick) the aim would be to use weld times between 5 and 20 cycles.

f) The pressure on the electrodes is maintained for a hold or forging time while the weld solidifies.

g) The pressure is finally released, the electrodes are moved apart, and the work is removed from the machine.

*Features of a resistance spot weld*
The weld which is produced by the sequence described contains three distinctive features (fig. 11.4). The central region, where the interface has been melted, has a cast structure typical of fusion welds, with columnar grains which meet at the line of the original interface. Surrounding the weld nugget is a heat-affected zone, showing that the parent metal has undergone a heating-and-cooling cycle. The outer surfaces of the sheets show indentations resulting from the pressure of the electrode tips. The reduction in thickness of the sheet at this point should not be more than 10% under normal conditions.

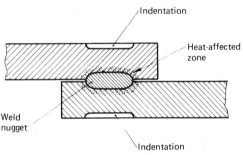

**Fig. 11.4** Features of a spot weld

*Controlling the welding current*
The easiest way of altering the current is to adjust the voltage across the electrodes.

According to Ohm's law, $I = V/R$. As $R$ is fixed by the force applied to the electrodes and the surface condition of the metal, any change in $V$ is reflected by a corresponding alteration in the current $I$. The voltage can be adjusted by changing the turns ratio of the transformer, and most resistance-welding machines are fitted with primary taps. Design considerations preclude the use of a large number of taps, and this method of adjustment can only give a coarse selection of current. In the simplest machines, where the current is switched on and off by an electromagnetic contactor in the main input line, this is the only means available to control current, and any finer setting of heat input must be achieved by adjusting weld time.

Where precise control of the welding sequence is needed, a more accurate method of switching must be used. In chapter 4 we discussed the use of thyristors in power-supply units for arc welding and we noted that, in

170

addition to their role in rectifying the alternating current, they could also be made to switch the current on at a selected point in each half cycle. A thyristor will not conduct electricity until it has been fired by applying a voltage to its control grid. There is virtually no delay in allowing current to flow in response to the control signal, so the thyristor offers very fast switching.

We must not forget that the thyristor still rectifies the current even when used as a switch. An alternating current is used in resistance welding, so it is necessary to have two sets of thyristors acting in opposite directions. In fig. 11.5, thyristor A looks after the positive half cycles, while B conducts only when the current is negative. By using a timer with a high degree of accuracy to supply the control voltage, weld times can be set precisely.

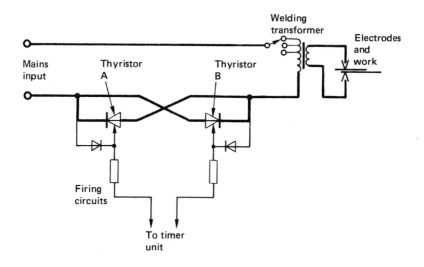

**Fig. 11.5** Thyristors installed in the primary circuit of a resistance welder to control current

In older units, ignitrons are used to control current flow (fig. 11.6). These are glass tubes filled with mercury vapour which conduct electricity only when a voltage is applied to the igniter.

A very important facility which follows from the use of thyristors for switching is the ability to control the level of current in the weld circuit. This is frequently called 'heat' control. The power in the weld sequence, and hence the total amount of heat developed, is proportional to the sum of the areas under the current waveform (fig. 11.7). If the start of each half cycle is delayed, a chopped sine wave is produced in which the area under the curve is reduced. The amount of power supplied is related to the proportion of the half cycle during which the current is flowing. Con-

171

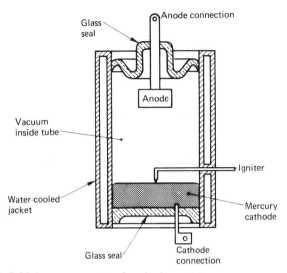

**Fig. 11.6** Main components of an ignitron tube

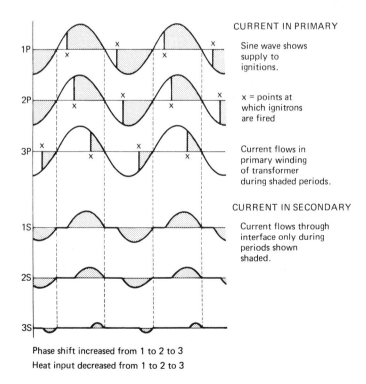

Phase shift increased from 1 to 2 to 3
Heat input decreased from 1 to 2 to 3

**Fig. 11.7** Effect of phase shift on current flow through the interface

trol of this can be achieved by adjusting the length of time between the beginning of each half cycle and the point at which the thyristor is fired. This technique is known as phase shifting. The amount of control which can be exercised in this way is limited, as there is a risk that the weld zone will cool between pulses if the delay time is long, and only a small proportion of each half cycle is used. However, by combining phase shifting with a limited number of primary tappings on the transformer, control over the complete current range of the machine can be made available.

*Machine rating*
It is not very helpful to classify or rate machines by quoting the current flowing in the secondary circuit, as we do not normally know the voltage. At the same time, measuring currents of the order of 2000 to 40 000 A flowing for only fractions of a second poses considerable problems and requires expensive equipment. A more meaningful approach is to quote the maximum amount of power available for conversion to heat at the interface of the sheets being welded. This is commonly expressed in kVA, thus a machine rated at 50 kVA draws 113 A from a 440 V mains supply. Ignoring losses in the transformer and connections, this amount of power is available at the electrodes. The current it produces in the secondary circuit depends, of course, on the voltage provided by a given tapping and on the interface resistance.

There will in fact be some losses in the system. These cause the temperature of the transformer to rise and, if current is drawn for long periods, the insulation could be damaged. To some extent this risk can be minimised by water cooling, but, to give a greater margin of safety, the kVA rating quoted for a machine specifies the power that can be drawn for thirty seconds out of each minute, thus allowing a cooling period.

*Current shunting in spot welds*
In practical welding situations, a further factor which affects the choice of current level is the spacing between neighbouring spot welds. When the first weld is made, all the current flows from one electrode to the other by the shortest and most direct route, i.e. along the extended axis of the electrode. With subsequent welds, the metallic bridge provided by the first spot weld offers an alternative low-resistance path and some of the current is diverted or shunted along this route (fig. 11.8). The total current is shared between the new site and the existing spot weld. The relative proportions depend mainly on the distance between the two points at the interface. Invariably the second weld is smaller for the same nominal current.

*Resistance welding high-thermal-conductivity metals*
The single-phase a.c. supplies considered so far work well for metals such as low-carbon steel, stainless steel, and nickel alloys. Heat can be developed rapidly at the interface, and melting can be achieved before significant amounts of heat have flowed into the surrounding areas. Metals such as aluminium which have a high thermal conductivity and a low body resistance

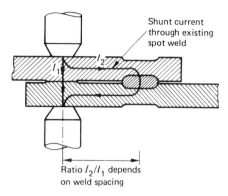

Fig. 11.8 Current shunting by an adjacent spot weld

require very large currents to compensate for the heat losses into the sheets. With thin material, this requirement can be met by using large-capacity machines rated at about 100 kVA. Where the problem is not readily solved by increasing the transformer size, three-phase transformers are used in conjunction with banks of thyristors to provide a suitable power level in the secondary circuit.

## 11.2 Practical resistance-welding systems

The principles of fusion by resistance heating which we have discussed in the preceding pages can be used in a variety of ways to weld lap joints in sheet material. However, in terms of industrial practice, only three systems have found wide application:

a) spot welding,
b) seam welding,
c) projection welding.

### Resistance spot welding

In many respects, resistance spot welding is the simplest of the three systems. The current flow is concentrated into a small area at the interface, and the result is a circular region of bonding.

The machines used for spot welding can be either stationary or portable. Stationary or pedestal welders range from small units for welding thin sheet to large machines having high-current outputs. They differ principally in the way in which the force is applied to the electrodes. With simple small machines a foot-operated pedal is linked to a pivoted arm which holds the upper electrode (fig. 11.9). Springs and weights are used to assist the operator and to increase the force applied to the electrodes. These units are often referred to as rocker-arm machines. A major disadvantage of this design is that the upper electrode is moved through an arc. This makes it difficult to align the electrodes accurately, and uneven wear often occurs at the tip of the electrode, producing a misshapen weld.

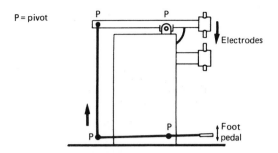

Fig. 11.9 Simplest form of foot-operated resistance spot welder

Greater force and more accurate alignment can be achieved with hydraulic or pneumatic cylinders (fig. 11.10). These can be fitted to rocker-arm machines, but better results are achieved by direct coupling to the electrode, and most of the larger welders rely on this method of achieving pressure at the interface.

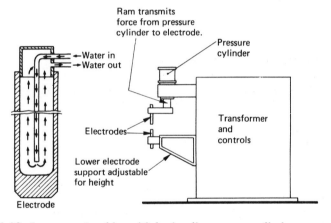

Fig. 11.10 Larger spot welder with hydraulic pressure cylinder

Portable units are widely used in the car industry. The transformer may be incorporated in the welding gun itself or be mounted remotely. Although the latter arrangement enables the gun to be lighter, it does mean that the cables to the electrode arms must be large, to avoid losses in the secondary circuit. Many of the smaller guns use a hand-operated lever to apply force to the electrodes (fig. 11.11). Bigger units, and especially those for repetitive work, are fitted with pneumatic cylinders to allow higher pressures to be used and to reduce operator fatigue.

In a number of box-like fabrications, there are situations where it is possible to bring the electrodes into contact with only one side of the joint.

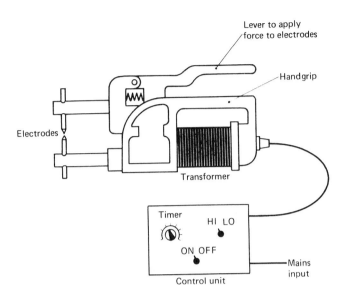

Fig. 11.11 Small portable gun with integral transformer

Generally this means that conventional resistance spot welding cannot be used, but, where a copper bar can be placed against the other side, a technique known as *series welding* may enable acceptable welds to be produced (fig. 11.12).

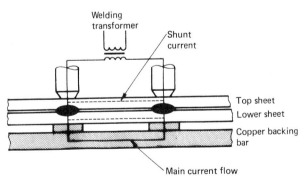

Fig. 11.12 Schematic diagram of series welding

With this method, two electrodes are positioned about 50 to 150 mm apart and force is applied, pressing the sheets against the copper backing bar. Ideally, the current flows from one electrode, through the sheets, along the backing bar, and back through the sheets to the second electrode. There

176

are thus two points on the interface where heating occurs and two spot welds are made simultaneously. In practical situations, some of the current flows along the top sheet, effectively reducing the amount of current flowing through the interface. Additional current must be supplied to compensate for this, and power requirements in series welding are therefore higher than in conventional spot welding.

### Resistance seam welding

The joints produced by spot welding cannot readily be made leak-tight, since they are bonded over only a small area of the interface. Admittedly, sealing compounds can be introduced between the overlapping sheets, but these are not always satisfactory, especially where flexing of the joint may occur, and their use increases the cost of the joining operation. A continuous bond or seam can be produced by overlapping the spot welds (fig. 11.13). After the first weld has been made, severe shunting occurs and the current must be raised to maintain the size of the welds. Accepting this limitation, satisfactory continuous seam welds can be made.

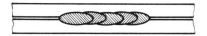

**Fig. 11.13** Overlapping spot welds used to produce a continuous bond at the interface

In the simplest form of resistance seam welding, the operator moves the sheet by a set amount after making each weld. The technique is often called *stitch welding*. With more automated units, the electrode movement and weld times are controlled according to a predetermined cycle and the operator is concerned solely with moving the component between electrodes during the periods when they are open, i.e. when there is a gap between them.

A further development is to use disc electrodes (fig. 11.14), one of which is driven by an electric motor. As the electrodes rotate, the work is moved between them and pulses of current are supplied. Each pulse lasts long enough to produce a spot weld, and the time interval is arranged so that the spots overlap by about 30 to 40% of their length along the seam. Continuous movement of the electrodes may not be possible with metals that are difficult to weld, e.g. Nimonic alloys used in aero engines. In these cases the electrode wheels will be moved through a given angle and then stopped while the welding takes place. They are then indexed round to the next point and a further weld is made. This allows the weld cycle to be timed independently of the speed of rotation of the electrodes, thus giving more flexibility in choosing the optimum welding conditions.

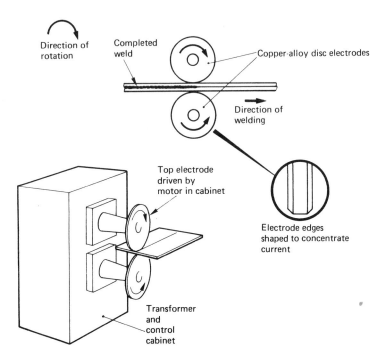

**Fig. 11.14** Seam welding

*Projection welding*

In both resistance spot and seam welding, current concentration or localisation at the interface is achieved by shaping the electrodes and restricting the area of the surface with which they are in contact. In projection welding, the weld is located by a projection raised on the surface of one of the sheets (fig. 11.15). When placed together, the sheets touch only at the point of the projection. Current flowing across the interface is concentrated through the projection, which heats up rapidly. As it becomes plastic and finally fuses, the projection collapses and a fused slug at the interface is formed. The result is similar to a spot weld.

Usually two or three projections are welded at the same time. The machines are essentially similar to those used for spot welding, but the electrodes are replaced by flat copper platens which exert a uniform pressure over the joint area. The choice of shape and size of projection is based on previous experience or experiment.

Projection welding is not often used for long lap joints. It finds much more useful applications in the joining of small attachments to sheet structures. It is widely used in the production of car bodies, domestic equipment, office furniture, and machine parts. For example, captive nuts can be made with small projections on one side which are used to attach them to a car-

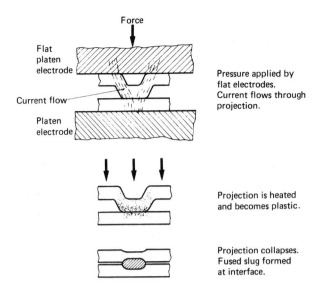

Pressure applied by
flat electrodes.
Current flows through
projection.

Projection is heated
and becomes plastic.

Projection collapses.
Fused slug formed
at interface.

**Fig. 11.15** Projection welding

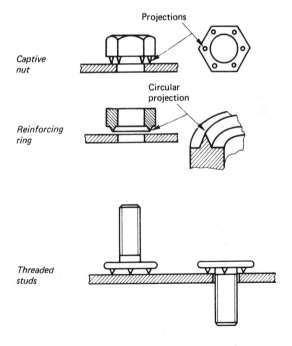

**Fig. 11.16** Examples of projection-welded details

chassis panel (fig. 11.16). Reinforcing rings are frequently projection-welded around holes in sheet-metal tanks. Threaded studs can be attached to backing bars by projection welding. A very important use of projection welding is in the manufacture of steel mesh for reinforcing concrete panels, making basket containers, and so on.

## 11.3 Quality control in resistance welding
Resistance welding poses novel problems in controlling quality. NDT techniques such as radiography and ultrasonics yield very little useful information about spot, seam, or projection welds. It would also be impossible to justify the cost of testing all the spot welds in the majority of sheet fabrications. On the other hand, the main variables in resistance welding, i.e. current, time, and presssure, are more readily monitored than are arc-welding parameters. If we can establish the conditions which give acceptable-quality welds, we can concentrate our attention on controlling the system so that they are reproduced for each weld.

With modern techniques, accurate timing of the weld cycle can be readily achieved. The use of thyristors also means that reasonably accurate setting of current levels is possible. Regular checking of the dimensions of the electrode tip coupled with the use of hydraulic or pneumatic cylinders provides good control of pressure. The only important variable left is the condition of the sheet material at the interface, and techniques must be established to ensure that a uniform or reproducible surface finish is used.

The extent to which these controls are exercised depends on the type of fabrication and the relative economics of the process. When welding flange joints in hot-air ducting, for example, it may be cheaper and just as acceptable to use simple equipment and increase the number of spot welds. On the other hand, the need for high quality in the welding of a flame tube in a jet engine justifies the inclusion of extensive quality-control measures to the extent of using meters to measure current and time.

It is also possible to monitor the quality and consistency by testing sample welds made at random intervals during the production run. Some typical tests are illustrated in fig. 11.17.

## 11.4 Flash welding
The principles of resistance heating to obtain fusion are applied quite differently in flash welding.

The two parts to be welded must be of uniform cross-section. Normally, the processs is used for joining bar, rod, or tubes, but there have been successful applications to sheet work. The parts are clamped with a small gap between the faces of the joint (fig. 11.18) and are connected to the output from a large transformer. There is a voltage of about 10 V across the gap so that a high current flows when the parts are moved into contact. As the surfaces are not smooth, contact is made at only a few points. The current flow is concentrated at these points, or bridges, which are rapidly melted by resistance heating. Immediately they are molten, the bridges are ruptured explosively, droplets of metal are ejected, and arcs are formed which heat

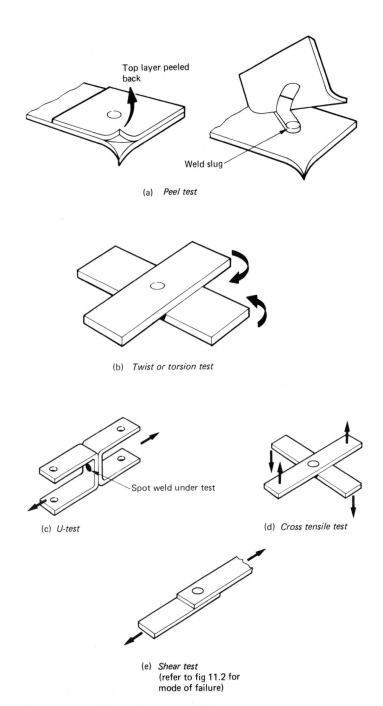

(a) *Peel test*

(b) *Twist or torsion test*

(c) *U-test*

Spot weld under test

(d) *Cross tensile test*

Top layer peeled back

Weld slug

(e) *Shear test*
(refer to fig 11.2 for
mode of failure)

**Fig. 11.17** Destructive tests on spot welds

181

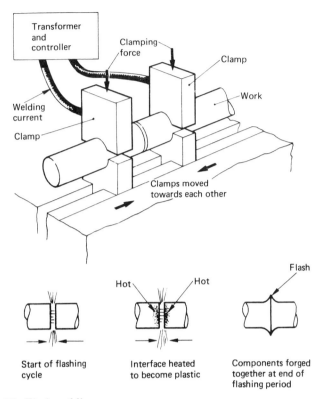

Fig. 11.18   Flash welding

the surfaces. There is now a gap between the parts. If the clamps are moved towards each other, contact is once again made and further bridges are established. The cycle repeats itself with more metal being ejected. This sequence is known as 'flashing' and is accompanied by showers of molten droplets over the surrounding areas. Flashing is an essential feature of the process, as during this period contaminants are removed with the molten droplets and the surfaces are heated to a uniform temperature. When these two tasks have been accomplished, a force is applied to the clamps, forging the parts together to give a solid-phase bond with any remaining molten metal being forced out to the surface. Here it forms a fin or flash around the joint. For some applications, such as pipe welding, it may be necessary to remove the flash, as it could obstruct the bore.

The equipment used for flash welding tends to be bulky. The clamps must exert sufficient pressure to prevent the components from slipping when the force is applied to bring them together. They must also restrain lateral movement, as correct alignment of the parts is essential if a uniform distribution of heat is to be achieved. At the same time, the clamps must be mounted on a rigid bed which does not flex or distort during the welding sequence.

Flash welding makes appreciable demands on the electrical power supply to a welding shop. Currents in excess of 100 000 A can flow across the interface, and power inputs of up to 200 kVA may be required. As the transformers used in flash welding are single-phase, an unbalanced load is placed on the 440 V three-phase supply used in most factories in the UK. Frequently this leads to problems which can be solved only by the installation of special transformers designed to distribute the load uniformly over the three phases.

Nevertheless, flash welding offers many attractions in the volume production of square or tubular components in a range of materials. It is particularly useful for the joining of non-ferrous metals such as Nimonic alloys and titanium. The absence of molten metal in the completed joints means that weld-metal cracking problems often associated with fusion welds are avoided, and that dissimilar metals can be joined. Some typical applications are the butt welding of continuous railway track and of boiler tubes, the manufacture of chain links, and the fabrication of metal window frames.

Usually, the joint is in the form of a butt with the axes of the components in line, but corner joints between sections and some 'T' joints are practical propositions. The main limitations are the maximum force which can be applied (the average requirement is about 70 $N/mm^2$ for low-carbon steel and up to four times this for heat-resistant metals such as Nimonic alloys) and the cross-sectional area which can be accommodated by the power supply available.

# 12    Welding and cutting with beams

In arc fusion welding, the width of the weld pool at the surface is usually between 5 and 10 mm, which means that the heat from the arc is spread over an area ranging from 40 to 180 mm$^2$, depending on the voltage, current, and travel speed. As the heat flows into the joint, the weld pool assumes its characteristic saucer-shaped profile.

A radically different approach to weld formation is used in a relatively modern group of processes of which the three most important are arc–plasma, electron beam, and laser welding. In each of these, the fusion and bonding are achieved by a technique known as 'keyholing', involving the use of a high-energy beam.

## 12.1 Keyholing
We can see how keyholing works by considering a beam directed at the surface of a metal plate (fig. 12.1). The beam has a small diameter so that it can be focused on to a spot on the plate surface about 1 or 2 mm in diameter. It also has a high kinetic energy, which it releases when it collides with a solid object. The amount of energy released needs to be in excess of 10 kW/mm$^2$ if a satisfactory weld is to be made.

At the point of impact, the temperature of the metal rises rapidly over a small area. The metal melts, and some vaporises. The molten metal is forced to one side, forming a crater exposing solid metal at the bottom of the small pool which has been formed. The beam can now impinge on the solid material and in doing so releases further energy. Melting of this newly uncovered metal occurs, another crater is formed, and the cycle is repeated until the beam has penetrated through the plate. At this stage there is a cylindrical cavity or *keyhole* extending through the thickness of the plate. The wall of this hole is covered by the molten metal which has been forced outwards from the centre-line of the beam. The metal is held in place by surface tension and by the pressure of metal vapour present in the hole.

We now have a situation in which the beam is travelling through the hole and emerging from the reverse side of the plate with little reduction in energy. If, however, we move the beam to one side, it comes into contact with the wall of the hole and releases its energy. The hole becomes temporarily elongated, heat is lost from the area vacated by the beam, and some of the metal on the trailing wall of the hole solidifies. Metal melted from the leading edge flows under surface tension round the hole to restore its circular cross-section. As the beam continues to move across the plate, this process of melting and solidification proceeds at a regular pace, keeping the shape of the hole uniform. At the completion of the traverse, a narrow band of cast

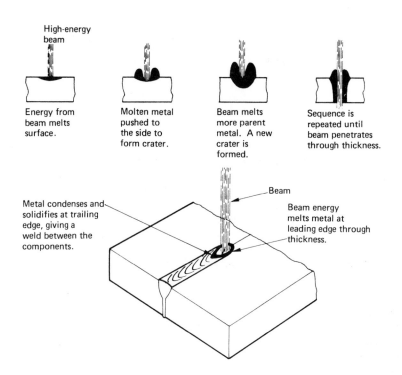

**Fig. 12.1** Principles of keyholing

metal has been produced through the thickness of the plate along the line of travel of the beam.

The sequence described is called keyholing. It offers a useful technique for welding since, by moving the beam along the line of a square-edge joint which has not more than a small gap between the mating faces, the molten metal coating the keyhole wall bridges the interface. Progressive solidification of the molten metal on the trailing wall bonds the two components of the joint, as in other fusion-welding systems.

Welds made by keyholing are characterised by their large depth-to-width ratio, which can be as high as 20:1 (fig. 12.2). In conventional arc fusion welding, where solidification proceeds from the fusion boundary to the centre, giving large grains, there would almost certainly be cracks in welds with such high depth-to-width ratios, assuming that they could be produced. Successful results are possible with keyholing because only thin layers of molten metal are involved in the solidification process at any given time. Small grains are formed, and the weld metal has good strength at high temperatures. The problem of ensuring an adequate supply of liquid metal to accommodate shrinkage is also made easier, since solidification proceeds along the line of travel, not vertically through the weld thickness.

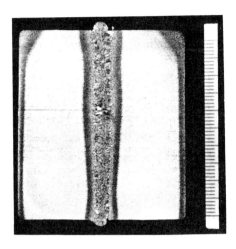

**Fig. 12.2** Cross-section of electron-beam weld (scale graduated in cm and 2mm divisions)

Keyholing offers two major advantages: uniform-penetration beads can readily be obtained and the parallel-sided weld gives uniform shrinkage with little or no distortion.

## 12.2 Cutting sequence
An essential feature of keyholing is the retention of the molten metal within the cylindrical hole. If this is removed or expelled from the joint, there can be no welding and a cavity remains after the beam has moved on. While such a situation is clearly undesirable in welding, we can turn it to our advantage as a cutting technique. A gas jet introduced concentric with the beam blows the molten metal out of the hole. If the beam and gas jet are moved across a solid plate, progressive melting and expulsion take place and a cut is produced. Unlike oxygen–fuel-gas cutting (see appendix B), the beam does not depend on a chemical reaction to achieve melting. The technique can be extended to a number of metals, such as stainless steel, aluminium, copper, nickel, and many others which are not suitable for oxygen–fuel-gas cutting.

Although a number of possible energy sources have been proposed for keyholing, only three systems are currently in use: arc–plasma, electron-beam, and laser.

## 12.3 Arc–plasma systems
In the context of our present discussion, the term plasma means a body of ionised gas. Plasma is produced when a gas is heated to a temperature which is high enough to dissociate it into positive ions and negatively charged

electrons. Energy is required to bring about this dissociation, and in welding this energy is supplied by the arc. The gas in the centre or column of the arc is dissociated at the temperatures involved, creating a plasma. As this gas flows away from the arc column, it reassociates to produce neutral atoms, giving up its energy in the form of heat.

The temperatures experienced in a tungsten arc used for TAGS welding are of the order of 11 000°C. The arc column is bell-shaped (fig. 12.3) and is free to move within the gas shield. If the arc is surrounded by a water-cooled copper tube, it is constrained and the temperature in the arc column increases to about 20 000°C. If the plasma gas is forced to travel through this constricted arc, it is heated to temperatures much higher than those in the conventional welding arc. It expands rapidly and issues from the constricting orifice as a high-temperature ionised gas jet. In other words, it is the type of high-energy beam required for keyholing.

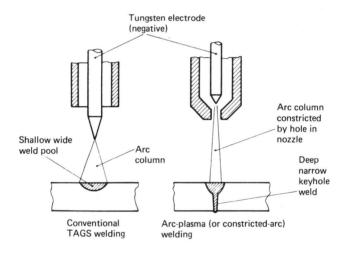

**Fig. 12.3** Arc–plasma welding

Using this, butt welds can be made in material of 3 to 15 mm thickness in one pass with currents of 100 to 300 A. Speeds are generally between 40 and 80% higher than for TAGS welding of the same joints. The process is not widely used, as its range is limited, but successful pipe-welding techniques have been developed to take advantage of the very good penetration-bead profiles that are characteristic of keyholing.

A low-current version, known as micro-plasma welding, has proved to be more successful. The system is operated at 0.1 to 10 A and uses nozzles much simpler than those required for arc–plasma welding. Micro-plasma welding is of particular value in the welding of metal less than 1 mm thick, as it offers stable operation at low currents – this is difficult to achieve with

TAGS welding. The welds are small and are more like those found in conventional arc welding. It seems unlikely that keyholing can operate effectively on very thin material when using plasma jets.

The major use of arc–plasma systems is in cutting. A hole is pierced through the plate, as with welding, but when the jet is traversed a cut is produced, with molten metal being blown out at the bottom of the gap or kerf (fig. 12.4). As the melting depends solely on the temperature of the jet, the composition of the metal being cut is not critical. Aluminium, copper, nickel, and other non-ferrous metals can all be cut satisfactorily.

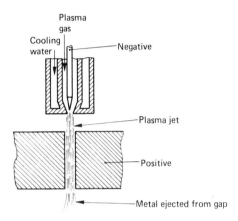

**Fig. 12.4**   Arc–plasma cutting

Cutting speeds are high. With currents between 100 and 300 A, aluminium can be cut at 0.07 to 3 m/min, depending on thickness. The maximum thickness is about 150 mm with the type of equipment generally available in the fabrication shop.

### 12.4 Electron-beam welding

An electron-beam welding machine has three main components (fig. 12.5):

a) a gun which produces a controlled beam of electrons,
b) a vacuum chamber with associated pumping equipment,
c) a unit which either traverses the beam along the joint line or moves the work under the gun.

**Electron gun**   Various types of gun have been used in commercial equipment for electron-beam welding, but in the main they are based on the same principles as the cathode-ray tubes and thermionic valves found in television sets. The source of electrons is a heated tungsten filament wire mounted in a cup-shaped electrode, which helps to control the flow of electrons (fig. 12.6).

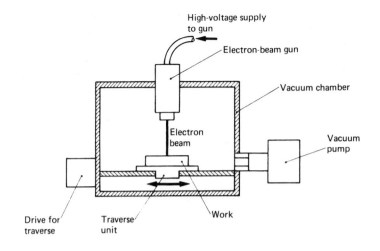

**Fig. 12.5** Main components of an electron-beam-welding machine

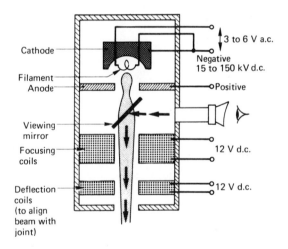

**Fig. 12.6** Main components of an electron-beam gun used for welding

The anode is usually in the form of a disc with a hole in the centre. Two ranges of voltage are used. In the low-voltage systems, the difference between the anode and the cathode is typically 15 to 30 kV, while a range from 70 to 150 kV is common for high-voltage machines. A number of operating advantages are claimed for both systems, but the resultant welds are essentially similar. At the voltages used in these guns, the electrons move rapidly from the cathode to the anode, where some are collected. Others pass through the hole in the centre of the anode and continue to travel until they meet a solid object, when they give up their kinetic energy in the form of heat.

The electrons which emerge from the hole in the anode are not necessarily travelling in the same direction, and the beam is usually divergent – in other words, it tries to take the form of a cone with the apex at the anode and the base on the surface of the work. If this were allowed to happen, the electrons would be spread over a relatively large area of the surface and it would not be possible to attain the high energy intensity which is essential for keyholing. The gun therefore contains magnetic coils which deflect the electrons towards the central longitudinal axis of the beam. By controlling the current flowing in the coils, and hence the strength of the magnetic field, the beam can in this way be focused to a diameter of about 0.5 mm.

With such a small diameter, it is of the utmost importance that the beam should be accurately aligned with the joint. Viewing windows can be fitted into the walls of the vacuum chamber. A much more satisfactory method is to sight along the axis of the beam by using a telescopic eyepiece and angled mirrors.

**Vacuum chamber**   If electrons are projected through air, they lose their energy in collision with oxygen and nitrogen atoms. Hence it is essential to operate in a vacuum at a pressure of $10^{-4}$ to $10^{-2}$ torr $(0.013$ to $1.3$ N/m$^2$) if the beam is to travel any distance.

The chamber must be large enough to accommodate the work and the gun. It must also be designed to protect personnel against radiation hazards. When the electron beam collides with the surface of the work, some of the energy released is converted to X-rays. These must not be allowed to escape, as they can harm workers in the vicinity, and they are contained by applying a lead lining to the chamber.

**Traversing the joint**   If the joint is small, it may be possible to complete the weld by moving the beam rather than the work. This has the advantage that mechanical traverse systems are not required – the beam can be deflected from its central line by magnetic coils. Current flowing in the coils is programmed to direct the beam along a predetermined path. The maximum deflection is relatively small, however, and for the majority of work the gun and the component must be moved relative to each other. In general, it is easier to move the work using a motor mounted outside the chamber. Because the traversing gear operates in a vacuum, dry lubricants must be used, as normal greases and oils evaporate and contaminate the chamber.

**Applications**   Electron-beam welding is particularly useful for the joining of highly reactive metals such as titanium and zirconium. These are kept free of contamination by the vacuum. The process is not limited to these materials, however, and many applications make use of the other unique attributes of this particular method of keyholing:

a) deep penetration, enabling one-pass butt welds to be made in plate up to 75 mm thick;

b) virtually uniform shrinkage about the neutral axis of the plate and narrow heat-affected zones, leading to very low levels of distortion – this allows fabrication of components without the need for post-weld machining;

c) removal of gases from the weld metal by virtue of the vacuum – for example, hydrogen is reduced to levels at which successful welding of hardenable steels can be achieved without the formation of heat-affected-zone cracks (see pages 103–7).

The high cost of vacuum chambers tends to limit the use of electron-beam welding to moderately sized components. In industrial practice up till now it has not been a major competitor to conventional fusion welding but has opened up new possibilities, especially in the manufacture of machinery. Gear clusters for transmission systems, intricate valve arrangements made from corrosion-resistant alloys, and pressure capsules are a few examples of components which can be welded in the pre-machined condition without significant loss of dimensional tolerances. Such items can be used in the as-welded condition without the need for the post-weld heat-treatment and machining operations often associated with arc fusion welding of similar work.

The component shown in fig. 12.7 is an example of a design which it would be difficult and expensive to machine from the solid. It can be readily made by machining three separate parts which are then electron-beam-welded. Only a small amount of machining is needed to remove the excess metal from the outer surfaces of the welds before the component is ready for service.

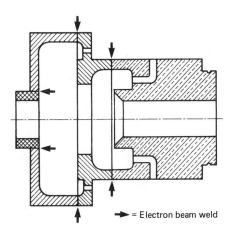

= Electron beam weld

**Fig. 12.7** Example of electron-beam-welded machined pump component

The electronics industry has made extensive use of electron-beam welding. The ability to locate small welds accurately is a considerable advantage in sealing capsules. Perhaps of more importance in many applications is the protection afforded to the materials used — welding in a vacuum means that there is no risk of oxidation which could impair the operation of semiconductors.

At the other end of the scale, advances in chamber and gun design have enabled large components in thick materials to be welded. The chamber illustrated in fig. 12.8 has a useful working volume of 7 m × 3.6 m × 3.6 m and is fitted with a 75 kW gun. It can be pumped down to an operating pressure of $5 \times 10^{-2}$ torr (7 $N/m^2$) in about 35 minutes and it can weld steel in thicknesses up to 150 mm. A 3 m long butt weld in this thickness would take about 20 minutes to complete.

Fig. 12.8    Large electron-beam-welding chamber at The Welding Institute

## 12.5  Laser systems

The heating effect of concentrated rays of light is well known. Sun rays focused on to a spot by a magnifying glass can make paper burn. By directing all the rays to one spot, the energy density is increased and the paper is raised to ignition temperature.

The amount of energy that can be harnessed in this way is limited. It is relatively easy to ignite a piece of paper, but the effect on a piece of steel would be negligible. Heat would quickly be conducted away, and the temperature would rise only a few degrees.

Visible light consists of a number of radiations of different wavelengths. The waves travel in random directions and are not in phase. Before we can achieve the high energy densities needed to melt metal, the radiations must be converted to the same wavelength and brought into phase using a *laser*.

The name 'laser' is an abbreviation of *L*ight *A*mplification by *S*timulated *E*mission of *R*adiation. In a simple laser (fig. 12.9), flashes or pulses of white light are directed into a YAG crystal. (YAG is an abbreviation of yttrium aluminium garnet.) The YAG crystal absorbs the energy and converts it into a single-wavelength beam of infra-red light with a diameter of a few millimetres.

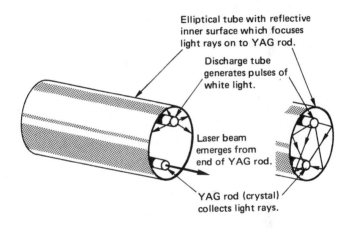

Elliptical tube with reflective inner surface which focuses light rays on to YAG rod.

Discharge tube generates pulses of white light.

Laser beam emerges from end of YAG rod.

YAG rod (crystal) collects light rays.

**Fig. 12.9** Principles of a YAG laser

The output of a YAG laser is limited by its inability to operate at high average powers. Current commercial equipment is limited to about 400 W average power, giving weld penetrations of about 1 mm. For deeper-penetration welding we need continuous operation at higher levels of power (up to 15 kW).

In commercial higher-power lasers used for welding, the YAG crystal is replaced by a tube filled with a mixture of carbon dioxide, nitrogen, and helium gases (fig. 12.10). A high voltage developed across electrodes placed inside the tube makes the gas fluoresce. The infra-red rays amplified within the discharge are reflected by mirrors at the ends of the tube, and the laser beam which emerges is focused to a point on the work. Output powers up to 20 kW are obtainable with $CO_2$ lasers.

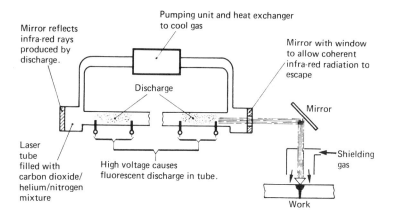

Fig. 12.10 Outline diagram of a $CO_2$ laser

## Laser welding

In many ways, laser welding is similar to electron-beam welding. As the beam hits the surface of the work, kinetic energy is released and a weld pool is formed.

At lower power levels, the weld pool is saucer-shaped and is similar to that of an arc weld. Use is made of low-powered lasers in the electronics industry for spot welding — typically for joining wires to flat surfaces and for securing edge joints.

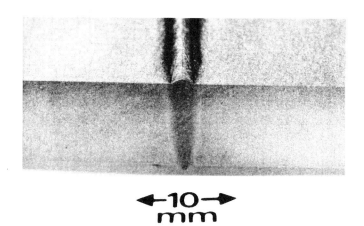

Fig. 12.11 Profile of a typical laser weld in 10 mm steel

With power levels above 1.5 kW, keyholing can take place and the welds take on the characteristic profile which, as we have seen, is associated with electron-beam welding (fig. 12.11). Welds of this type can be made in thicknesses up to 10 mm with commercially available lasers; but, above 6 mm, travel speeds are down to less than 1 m/min and stability problems in the weld pool can arise at these slow rates. One way of offsetting this is to use a wire-feed attachment to make a multipass weld, adding filler metal for the second pass (fig. 12.12).

**Fig. 12.12(a)** Wire-feed attachment for laser welding

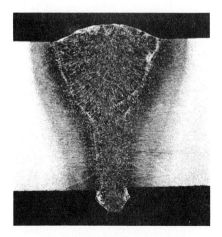

**Fig. 12.12(b)** Macrosection of a two-pass laser weld

Unlike with electron-beam welding, however, there is no need for the vacuum chamber with laser welding, as the laser beam travels readily through air. While this is a considerable advantage, it does mean that the weld is exposed to atmospheric contamination. The molten metal must be protected by a helium or argon gas shield in a similar way to TAGS welding.

The application of laser welding to industrial practice is still in its infancy. The electronics industry makes wide use of the technique, and there have been examples of successful laser welding of components for nuclear plant. More recently, manufacturers of cars and other domestic products have been incorporating lasers into their production lines.

Like electron-beam welding, a major attraction of laser techniques is the limitation of the heating effects. This is well illustrated in the manufacture of heart pacemakers. The electronic unit and its battery are encapsulated in a metal box or can. Because the battery and the electronic components would be damaged if they were raised to too high a temperature, soldered cans have to be made larger than would otherwise be necessary, to keep the components away from the heated area. The relatively small spread of heat in laser welding has meant that the size of the containers can be reduced appreciably, to the benefit of the patient.

### Laser cutting

Probably the widest use of high-power lasers in the fabrication and engineering industries is for cutting. This relies on keyholing to penetrate the thickness, but the molten metal is blown out of the hole by a gas jet. A nozzle is fitted concentric with the output from a $CO_2$ laser so that a gas jet can be directed at the work coaxial with the laser beam. The jet can be an inert gas, nitrogen, or, in the case of steel, oxygen. With the last of these, the oxygen reacts exothermically with the steel, giving additional heat as in oxygen–fuel-gas cutting (see appendix B).

Laser cutting is not confined to metals — wood, ceramics, plastics, cardboard, and synthetic clothing materials can all be cut by laser beams, often without the need of a gas jet.

# 13   Which process?

## 13.1  The range of factors influencing choice

In the preceeding chapters we have discussed many different welding processes. Each has its own particular characteristics and attractions. Some are widely used, others are known only in specialist circles. With so many alternatives on offer, what are the factors governing the choice of any one of them? This raises a series of questions: why do so many fabricators choose MMA to weld steel plate? what are the attributes of TAGS welding which make it more attractive than MMA to the aero-engine industry? why is resistance spot welding almost always the first choice of the car-body manufacturer? how in practice do we decide which process to use for any particular application?

It would be easy to say that the process chosen should provide the required quality at the lowest cost. While this must always be our aim, there are usually constraints which make our decision a compromise. We have earlier discussed the problems encountered in trying to define quality in welded joints. In making a choice between processes, we must consider quality in terms of the skill of the welders available to do the job. Similarly, availability of equipment contributes an important restraint – the volume of work may not justify investment in new plant or retraining labour, and existing welding sets are often used even though the apparent cost may be higher.

A further problem arises in determining the true cost of a process. At first sight it should not be difficult to work out the costs of a welding operation. Consumables and welding times can be readily measured. From the data obtained, a price per metre of welded joint can be deduced for each process and the results can be compared. But this does not give the complete picture. We must take into account the interaction between welding and other manufacturing operations. Some edge preparations cost more to cut than others. Welding processes and procedures which give significant amounts of distortion involve increased finishing costs in the correction of shape by straightening, and so on. Where the surface of the finished joint must be flat, the use of a technique which gives a smooth profile with a minimum of excess metal and spatter leads to savings in post-weld grinding or machining. In the fabrication of small machined components, a welding process which does not produce fume, spatter, etc. and which can be used alongside plant such as lathes and shapers offers marked savings in movement costs and production time. These few examples illustrate the need to consider welding as an integral part of an overall manufacturing operation. They also indicate some of the difficulties met in attempting to produce a true cost for welding.

The restraints we have identified vary from one situation to another. While recognising their existence, if we are to make sensible comparisons in general terms, we must have freedom of choice. In the following pages we need to assume that equipment and skilled labour are readily available and that ancillary fabricating operations such as cutting can be adjusted to suit each welding process.

## 13.2 Technical considerations

The first task is to ensure that the process chosen satisfies the technical requirements of the operation. The aspects which need to be considered are conveniently grouped under three headings: type of material, type of joint, and production requirements. A variety of questions can be asked about each of these aspects. The following list is not exhaustive – there are many more that could be asked in evaluating an application. The questions presented below aim to indicate the various factors which need to be considered.

### *Type of material*

a) Will the process remove oxides etc. from the joint and produce a fused bridge?

b) Can the weld metal be adequately protected against atmospheric contamination? As an example we can compare low-carbon steel with zirconium. With the former, all processes are satisfactory. Zirconium, however, is a very reactive metal and rapidly absorbs oxygen and nitrogen. Acceptable welds can be achieved with TAGS welding, but for the best results electron-beam welding is usually chosen.

c) Are flux residues likely to cause problems? One of the reasons why gas-shielded processes are preferred for aluminium is that fluxes in oxygen–fuel-gas and MMA welding corrode the metal if they are not completely removed.

d) Does the metal require special procedures to avoid the formation of cracks? Weld-metal hydrogen content in higher strength steels can be more easily controlled with gas-shielded processes than with MMA.

e) Are suitable consumables readily available? A major attraction of MMA welding is the ease with which fluxes can be formulated to meet special requirements.

f) Will the finished joint have acceptable properties? To a large extent this question can be rephrased to ask if adequate control can be exercised over heat input to guarantee reproducibility. Such a requirement is often an argument for the selection of a mechanised process.

g) Do the physical properties give rise to special considerations? Copper, with its high thermal conductivity, usually needs preheating to achieve fusion. The increased heat input in MAGS welding using an argon–helium shield enables a range of thicknesses in copper to be welded without preheat.

### *Type of joint*

a) Are butt or fillet welds required?

b) What is the position of welding? Joints in the vertical and overhead positions are rarely suitable for mechanised welding. On the other hand, many MMA electrodes are specifically designed for use in these positions.

c) Is there adequate access for the electrode, welding gun, or welding head?

d) What edge preparation is to be used? If large amounts of weld metal are needed to fill the joint, a high-deposition-rate process such as submerged-arc or spray-transfer MAGS welding is favoured, to keep weld times to a minimum.

e) Is an unsupported full-penetration root run required? This restricts the choice to low-current processes such as TAGS and dip-transfer MAGS, which give good weld-pool control. Alternatively, the use of a keyholing technique such as electron-beam welding may be justified.

### *Production requirements*

a) Are jigs, fixtures, or positioners required? These may alter access requirements.

b) Can the process be integrated with other fabricating operations?

c) What are the ventilation requirements?

d) Is there a power limitation which could preclude the use of high-current systems?

e) Can maintenance requirements be met?

### 13.3 Costs of the welding operation

Usually there is more than one process which meets the technical requirements of the job. This means that, in the absence of other constraints such as availability of labour, the final choice is decided by the relative costs of individual processes.

At this stage we should distinguish between arc and other processes. Resistance welding is akin to the many machine-operating sequences used in engineering and can be costed as such. It is also worth remembering that resistance spot welding is more often chosen in competition with mechanical fastening rather than with arc welding.

The essential components of the cost of an arc-welding operation can be summarised in a simplified formula:

$$C_w = L_w + L_a + O_h + C_c + P_m$$

where
$C_w$ = cost of welding operation
$L_w$ = cost for direct welding labour
$L_a$ = cost for associated labour
$O_h$ = overhead charges
$C_c$ = cost of consumables
$P_m$ = plant and maintenance cost

### *Direct labour*

It is relatively easy to identify the direct-labour element, since a welder is generally associated with one joint. There are some exceptions to this, but, even with the welding of large-diameter pipelines where two or three welders may work on the one joint, individual contributions can be readily recognised.

The biggest problem in assessing direct labour relates to the time spent actually welding, i.e. the arc time. In addition to depositing weld metal, the welder is involved in essential tasks such as setting the welding conditions, deslagging, electrode changing, and aligning the joint in mechanised welding. There are also other ways in which the welder can be occupied. For example, there may be periods of waiting for work to be delivered, or of moving from one place to another. Time must also be spent receiving instructions or discussing problems encountered during the welding operation. Electrodes and other consumables have to be drawn from the stores. Finally, there must be some relaxation time, as considerable concentration is required while the arc is being controlled in manual welding.

The welder is paid for all the time between clocking-on and the finish of the shift. In other words, all the non-arcing operations listed above form part of the total time allocated to the job and therefore are charged against the welding operation. For convenience these can be grouped as follows:

$$\text{total time} = \frac{\text{arcing}}{\text{time}} + \frac{\text{other constructive}}{\text{time}} + \frac{\text{waiting}}{\text{time}} + \frac{\text{idle}}{\text{time}}$$

To put these time elements into perspective, it is useful to think in terms of a duty cycle. This can be defined as the arcing time expressed as a percentage of the total time:

$$\text{duty cycle} = \frac{\text{arcing time}}{\text{total time}} \times 100\%$$

Since a welder is paid principally for the skill of fusing a joint and depositing weld metal, we obviously want as high a duty cycle as possible. There is a limit, however, as the welder not only needs relaxation times but must also perform the essential ancillary tasks. In addition, the layout of the welding shop often imposes limitations, with the result that the welder cannot achieve the theoretical maximum duty cycle. Similarly, the nature of the work has a marked effect on the distribution of the welder's time. Higher duty cycles can be maintained on the welding of long straight 'T' joints compared with attaching small fillet-welded lugs scattered around a box section.

In choosing a process, we must establish which one will give the best duty cycle for the job in hand, taking into consideration the related working environment. From what has been discussed, it is clear that there is no single answer to this problem. In general, however, we can say that higher duty cycles tend to favour continuous-wire systems such as MAGS and submerged-arc welding. In other words, these processes show to the best advantage on long joints, especially when they are welded in situations where the waiting and idle time can be kept to a minimum. The freedom from frequent electrode changes associated with continuous-wire systems enhances the duty cycle by reducing the 'other constructive time' element. At the same time, the faster travel speeds attainable reduce floor-to-floor time and thus increase productivity. By contrast, fast travel speeds and freedom from frequent electrode changes are of little value when depositing short weld runs, especially when the joints are far apart. These circumstances tend to favour MMA welding, where the increased manoeuvreability can help to raise an inherently low duty cycle.

### Associated labour

Mostly welders work by themselves, at least as far as the actual welding operation is concerned, but occasionally they need the help of another person to complete their task satisfactorily. In making an economic comparison of processes, the increased costs resulting from this additional labour must be taken into account.

Processes which reduce the extra cost to a minimum obviously have an inherent advantage. A good illustration of this is in the welding of higher strength steels, where preheat is specified with MMA welding electrodes. A change to MAGS or submerged-arc welding enables the preheat to be reduced or even eliminated, since the weld metal has a significantly lower hydrogen content (see page 104). This in itself is often sufficient to justify the selection of these two processes, but additional benefits such as fewer problems in electrode storage and handling and faster travel speeds also accrue.

### Overhead charges

A share of centralised expenditure must also be recovered as part of the welding costs. Functions such as both central and line management, quality control, design, stores and purchasing, sales, and general administration constitute an overhead for which production must pay. Various methods of recovering this overhead are used. One of the most common is to add to the labour costs a fixed percentage ranging from, say, 150 to 350%.

### Cost of consumables

Consumable costs are probably the easiest part of the assessment which we make when comparing processes. The volume of weld metal required can be readily calculated if we know the geometry of the joint. Data is available giving the recovery rates of electrode material, i.e. making allowances for losses by way of spatter, stub ends, and so on. From this, the quantity of electrodes required and hence the cost can be estimated. Similarly, once the arcing times have been established, the cost of electricity, shielding gas, fuel gases, and cooling water can be determined.

Replaceable items of equipment may be considered as part of the consumable costs. Opinions differ on this part of the cost comparison: some companies regard contact tips, cables, nozzles, and even MAGS guns as consumable items, while other allocate them to the cost of maintenance. Whichever approach is used, these replacement costs can be significant and must be taken into account in our process comparison.

201

*Machine and maintenance costs*

The last item in our formula relates to the capital cost of the welding plant and the charges made to maintain it in good condition. Mechanised plant is obviously much more expensive than a MMA welding set up. Actual prices could be quoted, but, since these change as a result of inflation, it is probably more helpful here to consider comparative costs, using MMA equipment as a base – Table 13.1.

Table 13.1 Comparative costs of welding equipment

| Process | Comparative cost |
|---------|------------------|
| MMA | 1.0 |
| TAGS | 2.5 |
| MAGS | 3.0 |
| Submerged-arc | 12.0 |
| Electro-slag | 20.0 |

The actual cost in terms of individual welding operations depends on the amount of use made of equipment and the depreciation policy adopted by the company's accountants. Very often, it reduces to only a small hourly charge compared with the cost of labour etc.

Maintenance costs are not easy to determine in advance, and usually an estimate is made based on previous experience. In the case of MMA welding, there is very little maintenance required, especially if transformer units are used. MAGS and mechanised welding systems do need regular maintenance, and an allowance must be made for this cost. The amount allocated varies from 5 to 15% of the capital cost of the plant, depending on whether items such as contact tips are regarded as consumables or as a maintenance charge.

## 13.4 Conclusions

It is difficult to summarise a subject as involved as process selection. Our discussions of the various aspects in the previous pages serve to emphasise the fact that there are no easy answers. In many instances our choice is a compromise, as we rarely have the freedom to adopt the ideal solution. Perhaps the most important point is that, whatever the constraints may be, a decision should not be made without first undertaking a careful analysis of all the factors involved. Nevertheless, certain key points do emerge:

a) For most of the metals used commercially in welded fabrications there is a wide choice of processes.
b) The processes currently available to the fabricator all give good-quality welds.
c) A major technical problem in choosing a process is the availability of suitable consumables.

d) Apart from welding low-carbon steel, the complexity of the best weld procedure is markedly influenced by the process, and this must be taken into consideration in our final choice.
e) Adequate skilled labour must be available. For manual processes, the craft skills are readily identifiable and are familar to those working in the industry. Gas-shielded and mechanised systems call for technician and supervisory skills which are not readily apparent and which may be overlooked in the initial comparison.
f) Cost analyses must be realistically related to the production environment and not based on a theoretical model.

Finally, it is worth noting that we are not necessarily setting out to select one process to do all the welding on a particular fabrication. Frequently, the most effective manufacture is achieved by using a mix of processes, even within a particular joint.

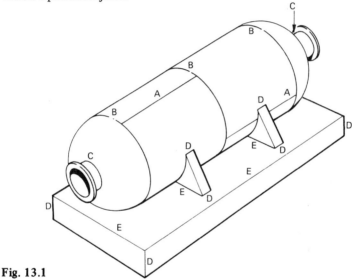

**Fig. 13.1**

Figure 13.1 shows a steel cylindrical storage vessel mounted on a base. Various solutions could be found to the problem of selecting the most suitable processes. One possible answer could be as follows.

*Joints A*   Submerged-arc welding with a removable flux backing would offer the fastest welding and would pose few problems in terms of access.
*Joints B*   The root runs could be deposited by MAGS welding, using a dip-transfer technique to achieve uniform root penetration. The joints could then be filled by submerged-arc or flux-cored MAGS welding with the cylinder rotated under the welding head. The latter process would have the advantage that joint tracking would be easier, since the welder could view the arc and correct alignment as necessary.

*Joints C*   MMA welding would probably be preferred for welding the nozzles into the domed ends, as access might be difficult and the weld runs are relatively short.

*Joints D*   Since these joints call for relatively short lengths of fillet or corner welds, it is most likely that MMA welding would be chosen.

*Joints E*   MAGS welding with a solid or flux-cored electrode could be used here, as the joints are long enough to enable the fabricator to benefit from the faster travel speeds.

These selections have been based largely on production considerations for shop fabrication. The picture would be very different if all the work had to be done on site with no positioning equipment, in which case the answer could be to use MMA welding for all joints. The selections would, of course, have to be costed for an actual component. In the end, we must be able to say that the welding process and procedure used has given the quality required at the lowest cost.

# Appendix A: safety in fusion welding

In common with other operations involved in the fabrication of metals, welding creates hazards which, if not identified, constitute a threat to the safety of both the welder and other workers in the vicinity. Many of these hazards are readily recognised because they form part of general engineering or fabrication work and are not specific to welding; for example, the handling of metal plates and sections, especially if they are hot, calls for care, as does the use of chipping and grinding tools. Both arc and oxy–acetylene welding present additional hazards, however.

## A1 Safety in arc welding

Arc welding poses special problems which originate from particular characteristics of the arc. These hazards can be grouped under three main headings (fig. A1):

a) electrical;
b) radiation
   i) visible light,
   ii) ultra-violet,
   iii) infra-red;
c) generation of fumes and gases.

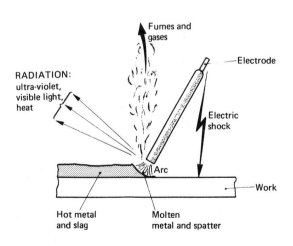

**Fig. A1**    Hazards in arc welding

## Electrical hazards

Electric shock occurs when a current flows through the human body and causes a muscular reaction. The level of current at which this happens varies from one person to another, but, in general, currents larger than 1 mA can be felt as an electric shock. At levels above 25 mA, the heart may cease to function.

With currents between 1 and 25 mA, two effects can be produced:

a) muscle spasm, with limbs 'twitching' or kicking, often leading to secondary injury by banging against sharp objects;

b) 'hold-on', a condition in which the muscles tighten the hand grip, making it impossible to release the live conductor.

Whenever electrical equipment is used, the operator must not come into contact with a voltage which is high enough to cause a current in excess of 1 mA to flow through the body.

It is difficult to avoid contact in welding, since the electrode and plate are both exposed and can be touched by the welder. Fortunately, most arcs in welding systems operate at between 10 and 40 V and are substantially safe in themselves. Dangers arise only when higher voltages are supplied to the work; we must, therefore, determine under what circumstances these can occur and devise methods of protecting the welder.

**Open-circuit voltage (o.c.v.)** The first and most obvious source of a higher voltage is a drooping-characteristic power-supply unit which is switched on but is not delivering current to the welding circuit. The units which are used for MMA, TAGS, automatic-arc, and submerged-arc welding need an open-circuit voltage (see section 4.4) significantly higher than the arc voltage, to assist arc striking and achieve the correct output characteristic. Some older transformer units were provided with o.c.v.'s of the order of 100 to 110 V. While these are much liked by welders because of the ease with which the arc can be struck, they are potentially dangerous since under adverse conditions severe shock can be experienced at 100 V a.c. In recent years, positive efforts have been made to design units having lower o.c.v.'s. In the case of d.c. power supplies, o.c.v.'s of the order of 60 V maximum can be used. When operating on a.c., however, some constraint is applied by the fact that often at least 80 V is required to ensure re-ignition of the arc each time the current goes through zero when the electrode changes from positive to negative and vice versa.

A useful approach to this problem is to protect the welder against exposure to the o.c.v. by the use of a low-voltage safety device (LVSD). Various equipments are available, but in the main they consist of a controlled contactor inserted into the welding lead. The contactor is in the open position when welding current is not flowing. The output terminal of the power supply is thus disconnected from the electrode, and only a low voltage exists between the holder and the work. As the electrode is touched to the work, a small current flows in the control unit which actuates the contactor, closing the contacts. The arc can now be struck, as o.c.v. is available at the end of the electrode. As long as the current flows, the contactor remains closed, but when the arc is extinguished the control circuit de-energises the contactor after a few seconds' delay. The welder is once again isolated from the power supply so that the electrode can be changed without risk of shock.

With MAGS equipment, electric-shock hazards are appreciably reduced, since o.c.v.'s are usually less than 50 V.

**Accidental connection of mains voltage to workpieces**  The welder can also be exposed to high voltages if a breakdown in insulation occurs either in the power-supply unit (not very common), in ancillary equipment, or in connecting cables with the result that the mains voltage is supplied direct to the workpiece or into the welding circuit.

Welding power-supply units are designed so that the primary and secondary circuits are completely isolated and the metal casings are connected to the mains earth. A provision for earthing the secondary circuit must be made externally to the set, and a correctly wired welding circuit should contain three leads (fig. A2):

a) a *welding lead*, connecting one output terminal on the power-supply unit to the electrode holder;
b) a *welding return*, connecting the other output terminal to the work;
c) an *earth lead*, connecting the work to an earth point.

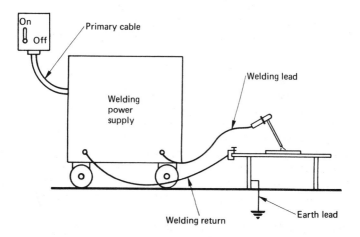

**Fig. A2**  Circuit for manual metal arc welding, showing earth connection

All three leads must be capable of carrying the highest welding current which will be used. It is also important that the earth lead should be connected to an independent earth point which is capable of taking the welding current — normally the mains earth is not rated high enough for this purpose.

## Check points for electrical hazards

Electrical safety requires systematic checking of details. Some of the important points to be checked are

a) that isolating switches are to hand so that the mains supply can be disconnected as quickly as possible;
b) that equipment is correctly rated and installed with adequate earthing;
c) that insulation on all cables is satisfactory;
d) that all connections are made with correctly rated components and are securely fastened;
e) that the welding lead, welding return, and earth lead are of adequate size for the currents being used;
f) that the earth lead is securely fastened to a satisfactory earth point.

Finally, every effort should be made to avoid welding in damp or wet conditions, since the danger from electric shock is much increased.


## Radiation from the arc

The first impression we get when looking at an arc is that it is a very bright light. This is because there is a high intensity of radiation which is within the visible range of the spectrum, i.e. of wavelengths between 0.4 and 0.75 $\mu$m. The intensity of light emitted depends on the current level and the presence of a flux.

The arc must be viewed only through a coloured filter (fitted into the welder's helmet) which reduces the light to an acceptable level. Filters are graded according to their density, and BS 679:1959, 'Filters for use during welding and similar industrial operations', recommends the most suitable grade for a particular application (Table A1).

The arc also emits ultra-violet and infra-red radiation with wavelengths less than 0.4 $\mu$m and greater than 0.7 $\mu$m respectively. These are outside the visible range.

Ultra-violet radiation is damaging to both eyes and skin. Exposure of the tissue of the eyeball to ultra-violet radiation will produce a condition known as 'arc-eye'. This is characterised by a painful gritty feeling under the eyelids, watering of the eyes, and an inability to tolerate light. Even a brief flash of ultra-violet radiation from a high-current welding arc is sufficient to cause severe arc-eye. The filters used must absorb as much as possible of the radiation with wavelengths of less than 0.4 $\mu$m. At the same time, attention must be paid to the effect of ultra-violet radiation on the skin. Small amounts of radiation simply result in a sun-tan effect, but prolonged exposure causes severe burning, and the welder should wear suitable protective clothing to cover any areas of skin which might be exposed to the arc.

While specific forms of cataract of the eye have been attributed to the effect of regular exposure to infra-red radiation, it appears that the risk is relatively small at the levels involved in welding. Nevertheless, as an extra precaution the filters specified by BS 679 will absorb infra-red radiation, thus ensuring that there is little danger of the welder's eyes being affected.

208

**Table A1** BS 679 recommended filters

| Filter | Process and current range (amperes) | | | |
|---|---|---|---|---|
| | MMA | TAGS | MAGS | Automatic metal arc |
| 8/EW | up to 100 | up to 15 | — | — |
| 9/EW | up to 100 | 15–75 | — | — |
| 10/EW | 100–300 | 75–100 | up to 200 | — |
| 11/EW | 100–300 | 100–200 | up to 200 | — |
| 12/EW | over 300 | 200–250 | over 200 | over 300 |
| 13/EW | over 300 | 250–300 | over 200 | over 300 |
| 14/EW | over 300 | 250–300 | over 200 | over 300 |

*Note* Where two or more shade numbers are recommended for a particular process and current range, the higher shade numbers should be used for welding in dark surroundings and the lower shade numbers for welding in bright daylight out of doors.

Ironically, the workers who probably run the greatest risk of arc-eye are not the welders but those who work in the vicinity of the welding operation. Although the intensity of radiation falls as we move from the arc, it can still be high enough to cause damage at some distance away. A further hazard is created by the reflection of the radiation from polished metal surfaces, white coats, and so on. Wherever possible, the welding operation should be surrounded by screens, and ancillary workers and other personnel should be warned not to view the arc with a naked eye.

### Check points for radiation hazards
To protect personnel against hazards arising from radiation from the arc, we should check that

a) the welder is using the correct grade of filter and a heat-absorbing filter is fitted where necessary;
b) the correct protective clothing is being worn;
c) the welding operation is screened as far as possible to prevent radiation from reaching other work groups;
d) ancillary personnel are warned of the hazards arising from viewing the arc with a naked eye;
e) first-aid treatment is immediately available for anyone who has been subjected to a 'flash' from the arc.

### Production of fume

At the temperatures existing in the arc, fume and dust particles are readily produced. These can be carried into the zone around the welder's face by convection currents of hot air rising from the arc. The extent to which these particles constitute a health hazard depends on their chemical composition. The maximum concentration to which a welder can be exposed must be specified for each chemical compound produced as fume by the arc. This concentration is known as the *occupational exposure limit* (OEL), and each year the Health and Safety Executive in the UK publishes a list of OEL's for substances commonly found in industrial atmospheres. These are not statutory limits but provide the basis for an assessment of the acceptability of an atmosphere.

The actual concentrations of pollutants that exist in the atmosphere in the area around a welding operation are measured by drawing a known volume of air through a filter. The increase in mass of the filter is determined, and the total amount of fume is calculated in $mg/m^3$. The fume is then analysed chemically to find the relative amounts of individual compounds, and the results are compared with the relevant OEL's.

Measurements are usually made in two areas (fig. A3). The first, known as *the breathing zone*, represents the atmosphere which the welder inhales and since it is closest to the arc, it normally has the highest fume concentration. The other is the region surrounding the welding operation and gives the

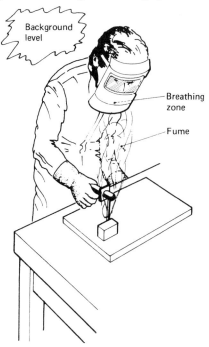

**Fig. A3**　Breathing zone and background level in arc welding

*background* level, which is of importance in considering whether it is safe for others to work in the area.

Measurements made in these two positions are influenced by four principal factors: the type of process, the current being used, the presence of surface coatings, and the type of site in which the welding is located. The last mentioned is most important, since to a large extent it determines whether the background level builds up. In open-air operations there is normally natural air movement which helps to disperse the fume. On the other hand, in a normal factory, although there may be some air movement in the shop, fumes can collect at ceiling height, giving a blanketing effect, while hot-air blowers used for heating may simply recycle the fume. The worst case is welding in a confined space, such as inside a tank, where the fume level can rapidly exceed the recommended limit.

Three approaches can be taken to the reduction and control of fume levels. Firstly, the atmosphere inside the welder's helmet can be removed and replenished by fresh air. Special helmets can be purchased for this purpose, and these probably provide the best method of protecting the welder in normal workshop situations, but it must be remembered that other people in the shop may still be exposed to a polluted atmosphere.

Secondly, the fume can be removed by placing an extractor duct in the vicinity of the arc (fig. A4). In theory, this presents the full answer to the problem since, if the fume is eliminated at source, the atmosphere in the work area will be free of pollutants. In practice it is difficult to achieve the full benefits, since the duct must be moved frequently during the operation, especially as it works at its maximum eficiency only over the central area.

**Fig. A4**   Local extraction of welding fumes

Nevertheless, local extraction sensibly used can make a significant contribution to the reduction in the level of fume in the atmosphere surrounding the welding operation.

Finally, overall ventilation of the shop by way of, for example, roof extractors can help to protect workers not directly connected with welding, such as fitters, crane drivers, and so on.

A problem frequently encountered in planning ventilation for welding shops is that of heat losses. As the number of air changes per hour required to reduce the concentration of fume to the OEL is usually large, systems which simply vent to outside atmosphere will draw large quantities of unheated air into the shop. While this may be acceptable in the summer months, at other periods in the year it can cause a dramatic increase in fuel bills. In these circumstances, recirculatory systems with filters to remove the dust and fume are worth considering, since the increased expenditure on installation may well be compensated for by savings in heating costs.

### Check points for fume hazards

It is important to remember that not all fume can be seen, and precautions should be taken even when there is no visible warning of the hazard. We should check that

a) material being welded does not produce compounds which have a low OEL;
b) surface coatings will not decompose to give poisonous gases or fume;
c) fluxes do not contain compounds with low OEL's;
d) local extraction can be used effectively and is not hindered by the shape of the work;
e) local-extraction systems do not exhaust into the shop in such a way that they produce a hazardous concentration at some point remote from the weld area – units incorporating filters are to be preferred;
f) background levels of fume, especially underneath the roofing, should be kept low by the use of general ventilation.

If there is any doubt, the atmosphere should be sampled and analysed.

### Gaseous contaminants

Gases are produced in the arc typically by decomposition of fluxes or by the effect of ultra-violet and infra-red radiation on air.

Ultra-violet light from the arc can cause oxygen in air to rearrange itself molecularly into ozone:

$$3O_2 \longrightarrow 2O_3$$

Ozone is very active chemically and, if inhaled, produces severe irritation of the lungs. This results in a loss of lung capacity and an inability to make even moderate physical exertions. In contact with solids, ozone reverts to oxygen, which means that it can be readily removed from air by passage through a filter. In this respect, ozone is unique among the gases found in welding atmospheres, since the others cannot be filtered out. Oxygen and nitrogen also react under the effects of heat and ultra-violet radiation to form nitrogen oxides which irritate the lung and, in high concentrations, can induce cyanosis (blue discolouration due to lack of oxygen in the blood) followed by death. A third toxic gas found in welding atmospheres is carbon monoxide, which is poisonous if inhaled in sufficiently large quantities.

Other gases met with in arc welding usually have a low toxicity, but it must be remembered that in TAGS and MAGS welding the gas supplied to the arc area will displace air. In an enclosed space there is a risk of asphyxiation as the air is diluted or replaced by the shielding gas. Air must be replenished by fresh supplies piped into the enclosed area at the same time as the diluted atmosphere is extracted (fig. A5).

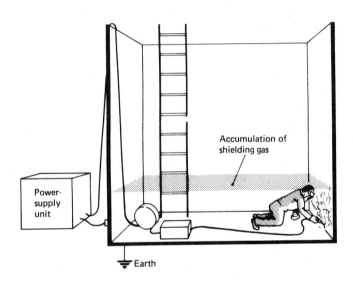

**Fig. A5**    Build-up of accumulation of shielding gas at the bottom of an enclosed space

*Check points for gaseous hazards*
Usually, the precautions taken to reduce the fume content in areas surrounding welding operations also remove gases produced by the arc. The main checks must therefore concentrate on ensuring that suitable ventilation is available. At the same time, it must be remembered that both argon and carbon dioxide, which are widely used in TAGS and MAGS welding, are heavier than air and may settle at floor level. An extractor duct placed at floor level can avoid the build up of a stagnant layer of asphyxiating gas. Finally, it is important to note that local extractors which vent into the shop will be discharging nitrogen oxides, and care should be taken to see that these do not pollute or contaminate other work areas.

**Vapours from degreasing fluids**  These require special mention since many are decomposed by the action of ultra-violet radiation to form phosgene or other poisonous gases. It is important that degreasing operations using solvents such as trichlorethylene are not carried out in the vicinity of welding operations, in case the vapour is drawn into the arc area. Similarly, work which has been degreased in this way should be completely free of solvent before welding operations begin. Machined components containing small cavities or crevices and castings with blow holes are particularly hazardous in this respect.

## A2 Safety in oxy–acetylene welding

Although the working principles of oxy–acetylene welding are similar to those of other fusion systems, it involves different considerations in terms of safety. These mainly arise from the different nature of the heat source.

Since heat is being generated by a chemical reaction, it is reasonable to assume that some of the products may present hazards. Firstly, oxygen from the surrounding air is used up in the burning of gases in the outer cone of the flame, and carbon dioxide is supplied to the weld area as a result of the combustion of acetylene. In an enclosed space, the air is diluted rapidly and the same precautions as those described for arc welding must be taken. Oxides of nitrogen are also produced and must be removed before their concentration can build up to the OEL.

The light output from an oxy–acetylene flame is much lower than that from an arc, and the levels of ultra-violet and infra-red radiation are relatively small. There is no need for a face shield, and it is sufficient for the welder to wear a pair of goggles fitted with a filter of the correct density. The recommendations in Table A2 are based on BS 679:1959.

**Table A2**  BS 679 recommended filters for oxy–acetylene welding

| Metal being welded | Recommended filter for | |
| --- | --- | --- |
| | Welding without flux | Welding with flux |
| Aluminium and alloys | 3/GW | 3/GWF |
| Bronze welding | 4/GW | 4/GWF |
| Copper and alloys, nickel and alloys, thin steel sheet | 5/GW | 5/GWF |
| Steel plate and thick sheet, pipe, etc. | 6/GW | 6/GWF |

Although we do not have to worry about the electrical hazards associated with arc welding, the heat source in oxy–acetylene welding does pose its own additional problems in that acetylene is an unstable gas and at critical concentrations can cause explosions. Great care must be taken to ensure that there are no leaks in fuel-gas and oxygen hoses and connections. In addition, oxygen must not feed back along the acetylene hose and vice versa, since this could lead to burning in the hoses which could spread to the cylinders. This feature, known as flash back, can be prevented by the use of check valves fitted to the torch. A shut-off valve or flash-back arrestor can also be fitted to the output from each regulator. The use of the correct shutting-down procedure also minimises the risk of a flash back.

### Check points for oxy–acetylene welding
Check that

a) adequate ventilation exists;
b) the welder is using the correct type of filter glass;
c) all connections in the gas-supply circuit are in good condition and free of leaks;
d) check valves and flash-back arrestors are correctly fitted;
e) oxygen and acetylene hoses are in good condition and of the correct colour:
     oxygen – blue,
     acetylene – red;
f) regulators and cylinders are connected correctly, with acetylene cylinders in an upright position;
g) all connections carrying oxygen are free of grease, since an explosion can occur if grease is in contact with pure oxygen.

### Storage of gases
Gases commonly used during welding are not toxic in themselves, and the usual problem is asphyxiation where concentrations of these gases are allowed to dilute the atmosphere to such an extent that there is insufficient oxygen for breathing. On the other hand, hazards are associated with the storage and handling of gases. Gases can be stored either in cylinders which are transported around the shop to be near to the appropriate welding site or in bulk containers from where they are distributed by means of pipes. The latter system creates fewer hazards and can make a positive contribution to improved safety.

Flammable gases such as hydrogen and acetylene are stored in red- or maroon-painted cylinders and have left-handed threads on connectors. Where hexagon nuts are used, the corners are notched to assist rapid identification (fig. A6).

Cylinders for non-flammable gases, such as oxygen and nitrogen, are painted black, while other gases such as carbon dioxide and argon may be identified by grey or blue paint. All cylinders used for non-flammable gases have right-hand threaded connectors fitted with unnotched hexagonal nuts.

Gas cylinders should always be stored carefully. Apart from acetylene and carbon dioxide, the gases inside are at pressures of the order of 170 bars, and care must be taken to prevent damage to the valves. A cylinder with a broken valve can behave like a jet-propelled missile. The pressure inside the

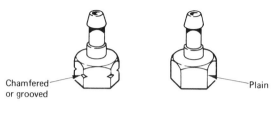

Chamfered or grooved

Plain

Left-hand thread
(combustible gas)

Right-hand thread
(non-combustible gas)

**Fig. A6**  Notched hexagon nut used for fuel-gas connections compared with plain nut for non-combustible gases

cylinders must not be allowed to exceed the stated maximum. This means that the cylinders must be stored in a cool place and must not be used close to an arc or a gas flame.

A number of recommendations regarding the storage and handling of industrial gases have been made over the years, and advice is readily available from the supplier of the gas.

# Appendix B: thermal-cutting systems

A number of thermal-cutting systems have been developed in which metal is melted along the cutting line and is blown out of the cut by a gas jet. Some of these have been widely used while others have found only a very restricted application.

In fabrication, the most commonly used techniques for thermal cutting are:

oxygen—fuel-gas,
oxygen—arc,
air—arc,
arc—plasma (see chapter 12).

## B1  Oxygen—fuel-gas cutting

Undoubtedly the most frequently used thermal process for the preparation of steel plate is oxygen—fuel-gas cutting.

To produce a cut, the steel is first heated by an oxygen—fuel-gas flame until it has reached the temperature at which the iron ignites when it is brought into contact with an atmosphere of high-purity oxygen. This is known as the oxygen-ignition temperature and is about $950^\circ$C. A jet of pure oxygen is directed at the heated area (fig. B1), 'burning' the steel in its path. Heat is released which melts surrounding metal and pierces a hole in the plate. If the nozzle is moved along the cutting line, so that the steel is progressively heated by the preheat flame and then oxidised by the oxygen jet, a cut is produced. The gap or kerf created by the cutting jet can be kept constant to within $\pm 0.5$ mm on plates up to 50 mm thick.

A cut of high quality can be obtained if the correct conditions are used and the speed of travel of the cutting jet is uniform. The latter is difficult to achieve if the torch is hand-held, although mechanical aids are available. In general, machine cutting in which the nozzle is traversed mechanically along the cutting line gives a higher quality of cut.

Typical travel speeds for oxy—acetylene cutting are given in Table B1.

### The chemistry of oxygen—fuel-gas cutting

The key reactions in oxygen—fuel-gas cutting are those between iron and oxygen since, not only do they convert the steel to an oxide slag for easy removal from the kerf, they also provide heat to keep the reaction going along the length of the cut. The two important reactions are

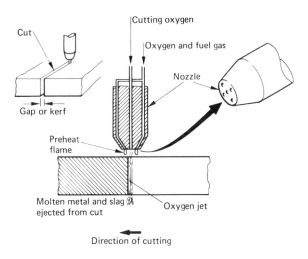

**Fig. B1(a)** Principles of oxygen–fuel-gas cutting

**Fig. B1(b)** Oxygen–fuel-gas cutting of steel plate

$$3\,Fe + 2O_2 \longrightarrow Fe_3O_4 + 1100\ kJ$$
$$Fe + O \longrightarrow FeO + 270\ kJ$$

As the cutting jet meets the hot steel at the top of the cut, some of the oxygen reacts with the metal. Heat is released, and a molten iron-oxide

**Table B1** Typical travel speeds for oxy—acetylene cutting

| Plate thickness (mm) | Travel speed (mm/min) | |
|---|---|---|
| | Square edge | Bevel edge |
| 6 | 420 | 370 |
| 12 | 360 | 320 |
| 25 | 250 | 220 |
| 50 | 180 | 170 |
| 100 | 160 | 150 |

slag is produced. The heat also melts some of the metal beneath the cutting face, and molten steel is mixed with the iron oxide. (Slag analyses show that about 30% of the iron in the material ejected from the bottom of the cut is not oxidised and hence was removed as molten steel.)

The slag produced by these reactions is blown down the cut by the jet. Heat in the slag raises the temperature of the next part of the cut, which is then brought into contact with the oxygen as the slag continues to be blown away. Further oxidation of the slag occurs, and more heat is generated. In this way the reaction proceeds through the thickness of the metal. Heat liberated at the top is carried down the cut, ensuring that the metal is always at the oxygen-ignition temperature when it is progressively exposed to the cutting jet. The only area which is not heated in this way is the top surface of the plate, since the heat generated there is carried down into the kerf. We must therefore use the preheat flame to keep the surface of the plate above the ignition temperature, otherwise the reactions do not take place and cutting ceases.

If the sequence of events we have just described is to proceed at a regular rate, it is imperative that the cutting oxygen must come into contact with 'clean' hot metal. In other words, there must not be any barrier which could prevent the oxygen from freely reaching the steel and thus slow down or even stop the reactions. How could such a barrier arise? So far we have only discussed reactions between iron and oxygen, but there are other alloying elements in steel which oxidise to form gases. Carbon produces a mixture of carbon monoxide and carbon dioxide, and sulphur is oxidised to sulphur dioxide. These gases are generated at the interface between the cutting oxygen and the steel, and must be swept away by the velocity of the oxygen jet. If they are allowed to collect, they form a stagnant boundary layer through which the oxygen diffuses only slowly or not at all (fig. B2). The amount of oxygen available at the surface of the steel is thus reduced, and insufficient heat is generated to keep the metal above the ignition temperature. The velocity of the cutting oxygen jet is therefore a critical factor in achieving cuts of acceptable quality.

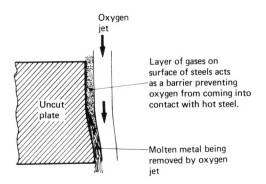

**Fig. B2** Formation of boundary layer in oxygen–fuel-gas cutting

Metallic oxides arising from the oxidation of alloying elements in the steel will not interfere with the cutting operation unless they increase the melting point or viscosity of the slag. A metal which behaves in this way is chromium, which is added as an essential element in stainless steel to obtain resistance to corrosion. The chromium oxide formed at the oxygen–metal interface raises the melting point of the slag and increases its viscosity. The slag is not readily removed from the reaction area, and an impervious boundary layer is established which prevents satisfactory cutting. The only practical solution to this problem is to add flux via the oxygen stream to reduce the melting point and viscosity of the slag. The quality of the cuts produced in this way is not of the same standard as that achieved on plain-carbon steels.

### The preheat flame

The preheat flame has two main functions: firstly, it raises the starting point to ignition temperature; secondly, it keeps the top layers of the steel hot so that the iron–oxygen reaction can begin as soon as the cutting jet impinges on the surface.

In theory, the choice of fuel gas is not critical, since any gas which when mixed with oxygen keeps the surface at the ignition temperature is acceptable: in practice, the choice of gas for the preheat flames plays an important role in the economics of the process. The gases most commonly used are acetylene, propane, butane, and proprietary gases based on hydrocarbon compounds.

Acetylene gives the highest flame temperature and the shortest times to raise the metal at the starting point to ignition temperature. It follows that, where the preheat time is an appreciable percentage of the total cutting time, acetylene can become a very attractive choice. On long cuts, on the other hand, the storage, availability, and cost of the fuel gas and the quantity of oxygen required will be more important, and gases such as propane and butane may be preferred. There is no single answer to such a choice, and each application must be evaluated separately.

A secondary function of the preheat flame is to burn off contaminants such as paint and grease from the surface of the plate ahead of the cutting stream. It can only cope with small amounts, however, and badly contaminated surfaces must be cleaned before the cutting operation is started.

### *Effect of oxygen–fuel-gas cutting on parent-metal structure*
Since gas cutting is basically a thermal process, a heat-affected zone is produced in the parent plate adjacent to the cut edge. In steels with very low carbon contents this is of little significance, since at the cooling rates experienced the steel is unlikely to harden. With higher carbon contents, bainitic and martensitic structures may be produced in the steel and the cut surface becomes hard. Provided this hardened zone does not contain cracks, the edge is acceptable for most welding applications. The extent of the heat-affected zone (fig. B3) depends principally on the plate thickness but also to some extent on the type of cut, i.e. square or bevel. Where the hardening effect is considered detrimental, heat treatment can be applied to temper the cut edge.

### *Equipment for oxygen–fuel-gas cutting*
The equipment used for gas cutting is similar to that described for oxy–acetylene welding. A supply of fuel gas and oxygen is required, and either

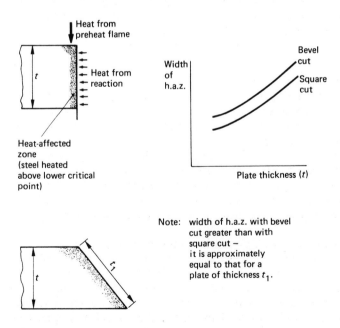

**Fig. B3**    Heat-affected zone in oxygen–fuel-gas cut edges

cylinder or bulk storage can be used. Since large quantities of oxygen may be needed during the cutting operation, it is more common in large fabricators' works to find bulk-storage tanks. Independent control of the preheat and cutting-oxygen supplies is essential.

One of the major problems which used to be encountered in oxygen–fuel-gas cutting was backfiring in the preheat flame, associated with burn-backs. The use of non-return valves and flash-back arrestors contributes to the safety of the operation, but the real answer lies in the careful design of torch and nozzle. As long as they are properly maintained and the correct pressures are used, modern torches give uniform burning of the oxygen–fuel-gas mix-ture at the top of the nozzle without the risk of backfiring. Badly worn nozzles or damaged nozzle seatings create situations under which this prob-lem can recur.

The nozzle and torch systems used for manual and mechanised gas cutting are essentially similar—the main difference between the two techniques lies in the better control of nozzle-to-plate distance and traverse speed which can be achieved by mechanisation. The simplest machines use a motorised cross-slide to move the cutting head across the plate. Slightly more complex units contain a facility for movement in two directions. Coupled with a template, this provides a method of profile cutting (fig. B4). Large machines can have more than one cutting head, and each head can be fitted with two or three nozzles to cut double bevel-edge preparations, with or without a root face, in one operation. These heads are motorised and can be guided by a variety of methods – templates, photoelectric followers, and numerical-control systems are probably the most popular techniques in current practice.

**Fig. B4**    Typical profile-cutting machine

## B2 Oxygen–arc cutting

The principles of oxygen cutting can be applied to arc cutting. Since the preheat flame in oxygen–fuel-gas cutting is principally used to heat the metal to the ignition temperature and then maintain it at this level, it can be replaced by other methods of heating, e.g. an arc.

In oxygen–arc cutting, a hollow electrode which has a flux covering is used. The electrode is connected to either a d.c. or an a.c. power-supply unit, although d.c. with electrode negative is usually preferred since it tends to give faster cutting speeds. The arc is struck at the edge of the plate, and the metal is rapidly heated to the ignition temperature. Oxygen is supplied to the hole in the electrode at a pressure of about 5 bars. The cutting jet thus produced impinges on the hot metal and an iron–oxygen reaction takes place as in flame cutting (fig. B5). The electrode is moved along the plate, with the outside edge of the flux covering in contact with the surface. Since the end of the electrode burns in the form of a cone, the arc length is thus kept constant. The travel speed is adjusted to give a uniform kerf width.

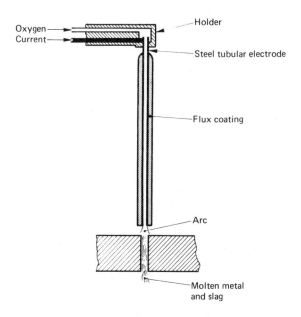

**Fig. B5** Principles of oxygen–arc cutting

The flux covering assists in cutting steels containing alloying elements which would otherwise retard or stop the exothermic reaction between iron and oxygen. Although cuts can be achieved on oxidation-resistant metals, the mechanism is quite different, since the arc provides the heat for melting and the oxygen jet is used to blow the molten metal out of the kerf.

Two sizes of electrode are commonly used: 5 and 7 mm diameter with holes which are 1.6 and 2.5 mm diameter respectively. The electrodes are usually 450 mm long, and the length that can be cut with one electrode depends on the thickness of the material.

### Data for oxygen—arc cutting of low-carbon steel

Table B2 gives typical data for oxygen—arc cutting of low-carbon steel; cutting speeds for those metals where the heat is supplied principally by the arc and not by an exothermic reaction are appreciably slower. For example, the speed for 25 mm stainless steel or Monel plate would be 4 m/h, while with bronze of the same thickness the figure would be 5 m/h compared with about 30 m/h for low-carbon steel.

**Table B2**  Data for oxygen—arc cutting of low-carbon steel

| Electrode size (mm) | Thickness of metal (mm) | Current (amperes) | Cutting speed (m/h) | Length of cut per electrode metre |
|---|---|---|---|---|
| 5 | 8 | 120 | 57 | 1.14 |
| 5 | 19 | 125 | 38 | 0.74 |
| 5 | 38 | 135 | 26 | 0.58 |
| 7 | 50 | 190 | 26 | 0.61 |
| 7 | 75 | 230 | 17 | 0.42 |

In general, the surface finish of the cut is rougher than that achieved with flame cutting or arc—plasma cutting. The process can be used in a mechanised form, but most applications involve manual operation. Guide bars can be readily used, since the insulating properties of the flux covering prevent shorting out when the electrode touches the bar. The depth to which the heat penetrates is smaller than in flame cutting, and higher cooling rates can occur, with a greater attendant risk of hardening in carbon and alloy steels.

## B3  Air—arc cutting and gouging

The air—arc technique for cutting is significantly different from flame and oxygen—arc cutting. All the heat is provided by an arc, and a jet of compressed air is used to blow the metal out of the cut (fig. B6).

The electrode material in air—arc cutting is a combination of carbon and graphite, and usually it is coated with copper. The carbon electrode is held in a specially designed holder containing holes through which jets of compressed

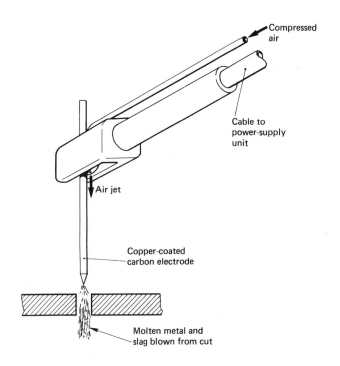

**Fig. B6**    Principles of air–arc cutting

air blow along the electrode. The electrode is connected to the positive terminal of a d.c. supply unit, and the arc is struck between the electrode and the surface of the parent metal. A molten pool is established, and the metal is blown from beneath the arc by the compressed air. Further melting takes place followed by removal of the metal until the cut has penetrated the thickness. The arc is then traversed along the line of the cut. The surface finish is not good compared with that produced by flame cutting, but, since the heat is supplied by the arc, both non-ferrous and ferrous metals can be cut.

The air–arc system has only a limited application in cutting. More commonly it is used as a gouging tool to prepare the reverse side of a weld for the deposition of a sealing run or to remove metal which contains defects. In the hands of a skilled operator, grooves can be produced which require little preparation before being filled with weld metal.

# Appendix C: electrode specification

**British Standard specification for electrodes used in MMA welding of carbon and carbon–manganese steels (BS 639:1976)**
In Great Britain, the specification of electrodes for MMA welding is covered by British Standard BS 639:1976, 'Covered electrodes for the manual metal arc welding of carbon and carbon–manganese steels'. This standard is based on the recommendations of the International Organisation for Standardisation (ISO) given in document ISO 2560. It provides both the manufacturers and users of the electrodes with a comprehensive statement of the minimum requirement for the various types of MMA electrodes used for the welding of carbon and carbon–manganese steels.

In addition to specifying properties, the standard also details a system by which electrodes can be given a code number. This makes it easier to identify the different types without using trade names. Thus a typical general-purpose electrode might be designated E4321R13, while an electrode for welding higher-strength steel, when the weld pool must have a low hydrogen content, might have the reference E5133B20H. These specification numbers provide two sets of information: the first four digits indicate the minimum mechanical properties which can be obtained in the weld; the remaining part of the code provides information about the operating characteristics.

## Mechanical properties
The standard specifies the minimum yield stress, tensile strength, ductility, and Charpy impact strength to be expected from each type of electrode. In most cases the manufacturer designs the electrode so that the properties are better than this minimum requirement.

The test piece used to check that the properties are satisfactory consists of a butt weld in 20 mm plate. The edges of the joint are bevelled, and a backing strip is used. The root gap is 16 mm, which is much larger than that used in normal practice – this is done to ensure that both the tensile test-piece and the notches of the Charpy specimens are entirely in weld metal and are not affected by the plate.

Two levels of tensile strength are specified. Electrodes giving weld-metal strengths within the range 430 to 550 $N/mm^2$ are designated E43, while E51 electrodes give 510 to 650 $N/mm^2$ (Table C1). The minimum yield stresses for these are 330 $N/mm^2$ and 360 $N/mm^2$ respectively. From the point of view of design, the yield stress is more important, and the electrode manufacturer usually supplies more specific data on expected weld-metal strengths as supplementary information to the code number.

Ductility is expressed as *percentage elongation* of the tensile test-piece. In the code number, this is linked with the Charpy *impact values* (see fig. 7.14), which should be obtainable in tests conducted at various specified temperatures. The second two-digit group of figures designates, therefore, the minimum values for both elongation and impact strength (Tables C2 and C3). At first sight these provide conflicting information. Let us consider an E51 33 electrode. The first '3' tells us that the minimum elongation is 20%, while the second '3' specifies that it should be 22%. Similarly, the two digits refer to Charpy values of 28 J and 47 J respectively at $-20^{\circ}$C. To understand this apparent anomaly, we must recognise that the information is based on two different methods of conducting the tests. The first digit refers to the method specified in ISO 2560; the second digit relates to the method which was used in earlier British Standards and thus provides continuity with the old system of classification. In both cases, however, the Charpy values are only quality-control acceptance levels and *cannot* be used in design calculations. The code number '33' simply tells us, therefore, that when tested at $-20^{\circ}$C the weld will have acceptable ductility and impact strength.

### Flux covering
The letter in the middle of the code number indicates the type of flux (Table C4). This aspect has been discussed in detail in chapter 5. It is sufficient here to draw attention to the fact that the operating characteristics of the electrode are influenced by the type of flux.

### Operating characteristics
Three important pieces of information about operating characteristics are also provided by the code number.

**Deposition rate,** i.e. the mass of deposited weld metal as a percentage of the core-wire mass, can be included where it exceeds 100%. This occurs when the flux covering contains iron powder which is melted and added to the fused core wire. Thus, with an E43 21 R12030 electrode the mass of the deposited metal is 20% more than the mass of core wire which has been melted by the arc.

**Positions of welding** for which the electrode is suitable are indicated by the next digit according to the code given in Table C5.

**Electrical conditions** (Table C6) relate to the type of power supply required. Three requirements are indicated:

i)   type of current, d.c. or a.c.;
ii)  electrode polarity (if d.c.);
iii) minimum open-circuit voltage required to achieve satisfactory striking of the arc.

227

## Typical specification

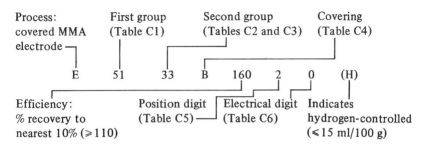

### Table C1 First group

| Electrode designation | Tensile strength (N/mm$^2$) | Minimum yield stress (N/mm$^2$) |
|---|---|---|
| E43 | 430–550 | 330 |
| E51 | 510–650 | 360 |

### Table C2 Second group, first digit

| First digit | Minimum elongation (%) E43 | Minimum elongation (%) E51 | Temperature for impact value of 28 J (°C) |
|---|---|---|---|
| 0 | Not specified | | Not specified |
| 1 | 20 | 18 | +20 |
| 2 | 22 | 18 | 0 |
| 3 | 24 | 20 | −20 |
| 4 | 24 | 20 | −30 |
| 5 | 24 | 20 | −40 |

**Table C3**  Second group, second digit

| Second digit | Minimum elongation (%) | | Impact properties | | |
| | | | Impact value (J) | | Temperature (°C) |
| | E43 | E51 | E43 | E51 | |
|---|---|---|---|---|---|
| 0 | Not specified | | Not specified | | |
| 1 | 22 | 22 | 47 | 47 | +20 |
| 2 | 22 | 22 | 47 | 47 | 0 |
| 3 | 22 | 22 | 47 | 47 | −20 |
| 4 | Not | 18 | Not | 41 | −30 |
| 6 | relevant | 18 | relevant | 47 | −50 |

**Table C4**  Covering

| | |
|---|---|
| A | Acid (iron oxide) |
| AR | Acid (rutile) |
| B | Basic |
| C | Cellulosic |
| O | Oxidising |
| R | Rutile (medium-coated) |
| RR | Rutile (heavy-coated) |
| S | Other types |

**Table C5**  Position digit

| | |
|---|---|
| 1 | All positions |
| 2 | All positions except vertical-down |
| 3 | Flat and, for fillet welds, horizontal – vertical |
| 4 | Flat |
| 5 | Flat; vertical-down; and, for fillet welds, horizontal – vertical |
| 9 | Any position or combination of positions not classified above |

**Table C6**  Electrical digit

| Code | Direct current<br>Recommended electrode polarity | Alternating current<br>Minimum open-circuit voltage (V) |
|------|--------------------------------------------------|---------------------------------------------------------|
| 0 | Polarity as recommended by manufacturer | Not suitable for use with a.c. |
| 1 | + or − | 50 |
| 2 | − | 50 |
| 3 | + | 50 |
| 4 | + or − | 70 |
| 5 | − | 70 |
| 6 | + | 70 |
| 7 | + or − | 90 |
| 8 | − | 90 |
| 9 | + | 90 |

# Index

ultrasonic flaw detection, 122–6

vacuum chamber, 190
vertical welding, 75, 78, 91, 154–8
visual inspection, 114–18

weaving, 75
weld
    definition, 11
    formation, 11
    nugget, 170
    pool, 20–2
welder approval, 128, 131

welding
    current, 72, 87, 93, 169
    position, 73–5
    procedure, 128–31
    speed, 43, 144
work-hardening, 102

X-rays, 118–21

YAG laser, 193

zirconiated electrode, 88